Sound-Power Flow

A practitioner's handbook for sound intensity

Sound-Power Flow

A practitioner's handbook for sound intensity

Robert Hickling

Sonometrics Inc., Huntington Woods, Michigan, USA

Morgan & Claypool Publishers

Rights & Permissions
To obtain permission to re-use copyrighted material from Morgan & Claypool Publishers, please contact info@morganclaypool.com.

ISBN   978-1-6817-4453-7 (ebook)
ISBN   978-1-6817-4452-0 (print)
ISBN   978-1-6817-4455-1 (mobi)

DOI   10.1088/978-1-6817-4453-7

Version: 20161201

IOP Concise Physics
ISSN 2053-2571 (online)
ISSN 2054-7307 (print)

A Morgan & Claypool publication as part of IOP Concise Physics
Published by Morgan & Claypool Publishers, 40 Oak Drive, San Rafael, CA, 94903 USA

IOP Publishing, Temple Circus, Temple Way, Bristol BS1 6HG, UK

This book is dedicated to the memory of Professor Milton S Plesset

Contents

Preface

There are three fundamental quantities in acoustics: sound pressure, sound particle velocity and sound intensity. This book is about sound intensity. However the term intensity does not clearly express the nature of the quantity and the term has occasionally been misused. The expression sound-power flow is used here instead. Sound-power flow is exactly that. It is a vector quantity representing the net flow of sound power in watts per square meter in the direction of sound propagation. Sound-power flow can be measured in fluids such as air and water, but it cannot generally be measured in solids. A secondary purpose of the book is to use linear units rather than logarithmic units, thus making computation in acoustics simpler and more accessible to advanced mathematics and computing. Another purpose of the book is to demonstrate the advantages and uses of acoustical sensing compared to other forms of sensing. The book is based, to a large extent, on work by the author and associates documented in internal reports by his group at General Motors, the University of Mississippi and Sonometrics. Most of this work subsequently was published in journals. In general, the purpose of the book is intended to be a guide for practitioners and research scientists in different areas of acoustical science.

The measurement of sound-power flow in fluids followed a path that originated in work by the author and his group, as documented in reports at the General Motors Research Labs in the 1970s. This included the development by J Y Chung of the cross-spectral method, of measuring a component of the sound-power flow vector, using two sound-pressure sensors. The book traces progress since that time.

The book presents computational approaches using standard mathematics and relates these to the measurement of sound-power flow in air and water. Sound-power flow cannot generally be measured in solids. However, it can be investigated by combining measurements at a fluid–solid interface with the corresponding computation of sound-power flow in the interior of the solid.

Related to measurement of sound-power flow in fluid media, applications described in the book include:

(a) Measuring the total sound power emitted by noise sources.
(b) Measuring sound-power flow in ducts.
(c) Directional hearing of humans and mammals.
(d) Use of the sound-power flow vector to determine the direction of a sound source.
(e) Echolocation.
(f) Use of the elastic response to incident sound of an object in water to determine the object's structure and nature.
(g) Uses of non-contact, focused, high frequency, pulse-echo ultrasonic probes.

After General Motors, work on sound-power flow was continued at the University of Mississippi by the author and his group. Accurate measurements were made of the sound power of a number of different mechanical noise sources. In water, measurements were made with a four-hydrophone vector probe. Sound-power flow and its visualization were computed in fluids and solids. Also, for the first time, uses were developed, for non-contact, pulse-echo, focused ultrasonic probes.

Subsequently work was continued at Sonometrics Inc. using small, sensitive electret microphones manufactured by Knowles Electronics Holdings, Inc. Accurate, inexpensive, four-microphone vector probes were developed for locating sound sources, together with two-microphone probes for scanning vibrating surfaces and for measuring sound power. This work had considerable help using illustrations by Rob Farmer.

The well-known predecessor of this book is the text Sound Intensity by Frank J Fahy of the Institute of Sound and Vibration Research at the University of Southampton in the UK. Needless to say, there are significant differences between this text and the present book. Another important predecessor is the four-volume Encyclopedia of Acoustics, edited by Malcolm J Crocker. The present book provides a more wide-ranging and somewhat different set of topics.

Robert Hickling
Sonometrics Inc.
Huntington Woods
Michigan, USA

Author biography

Robert Hickling

Robert Hickling's main area of research is engineering acoustics with particular focus on application of sound-intensity measurement, motor vehicle and environmental noise, and noise and overpressure due to airbags. Born in Bologna, Italy, he received an MA (Pure and Applied Mathematics) in 1954 from the University of St Andrews, Scotland, and a PhD in Engineering Science in 1963 from the California Institute of Technology. He has been an industry consultant for Delphi, Autoliv and Takata and is a fellow of the ASME and ASA. He was a Scientific Officer, for the Royal Naval Scientific Service, UK. He worked at the General Motors Research Laboratories, Michigan, USA. He held the post of Research Professor of Engineering Research at the University of Mississippi and is now a Professor Emeritus. He is the owner and president of Sonometrics Inc.

Chapter 1

Mathematics and measurement of sound-power flow in fluids

1.1 Introduction

In a fluid (i.e. gases or liquids) sound consists of compressional waves. In a homogeneous, non-viscous fluid, conservation of mass and momentum [1–3] are expressed using the following equations

$$\frac{\partial \rho}{\partial t} + \rho_0 \nabla \cdot \mathbf{v} = 0 \tag{1.1}$$

$$\rho_0 \left(\frac{\partial \mathbf{v}}{\partial t} \right) = -\nabla p \tag{1.2}$$

where ρ is the change in fluid density due to sound, ρ_0 is the undisturbed fluid density at atmospheric pressure $p_0 = 1.013 \times 10^5$ Pa, t is time, $\mathbf{v}$ is the sound particle velocity vector and p is sound pressure in Pascals. T_0 is temperature in degrees Kelvin (273.15 K) associated with ρ_0 and p_0. ∇ is the vector gradient. Vectors are represented by bold type. Equation (1.2) is called Euler's equation. Compressional sound waves in a gas are assumed to obey the adiabatic gas law

$$\frac{(p + p_0)}{p_0} = \left[\frac{(\rho + \rho_0)}{\rho_0} \right]^\gamma \tag{1.3}$$

where γ is the ratio of specific heats of the gas. The speed of sound c in a gas is given by

$$c = c_0 \sqrt{\frac{T}{T_0}} \tag{1.4}$$

doi:10.1088/978-1-6817-4453-7ch1
1-1

where T is the absolute temperature of the gas in degrees Kelvin and c_0 is given by

$$c_0^2 = \frac{\gamma p_0}{\rho_0} \tag{1.5}$$

Parameters for different gases are given in Appendix B and in [1, 3]. For example, air has $p_0 = 1.013 \times 10^5$ Pa and $T_0 = 273.15$ K: ρ_0 is 1.293 kg m^{-3}, γ is 1.402 and c_0 is 331.6 m s^{-1}.

Compressional waves in liquids obey a different equation of state

$$p - p_0 = \frac{B(\rho - \rho_0)}{\rho_0} \tag{1.6}$$

where B is the adiabatic bulk modulus given by

$$B = \rho_0 \frac{\partial p}{\partial \rho} \tag{1.7}$$

evaluated at $\rho = \rho_0$. Parameters for liquids are given in [3]. For example, fresh water has T_0: ρ_0 is 998 kg m^{-3}, B is 2.18×10^9 Pa, γ is 1.004 and c is 1481 m s^{-1}.

1.2 Average sound-power flow and the cross-spectral formulation

The instantaneous sound-power flow vector $\mathbf{I}$ in a fluid is derived from the linear equations of acoustics [2] as

$$\mathbf{I} = p\mathbf{v} \tag{1.8}$$

However it is not the instantaneous value that is of principal interest, it is the net or average sound-power flow. Average sound-power flow is expressed as

$$\langle p\mathbf{v} \rangle_{\text{avg}} = \frac{1}{t} \int_0^t p\mathbf{v} \, \mathrm{d}t \tag{1.9}$$

where the time period t can be of arbitrary duration. It is determined from sound pressure, using microphones in air and hydrophones in water.

Finding the velocity $\mathbf{v}$ from equations (1.1) and (1.2) is a first step in measuring average sound-power flow. Using finite-difference approximations based on Taylor series expansions, the pressure gradient in equation (1.2) is determined using the difference between two pressure measurements $p_1(t)$ and $p_2(t)$, divided by the separation distance d between the pressure sensors, i.e. $\frac{p_2(t) - p_1(t)}{d}$ at a point midway between the pressure sensors 1 and 2. The sound pressure at this point is $\frac{p_2(t) + p_1(t)}{2}$.

These approximations are valid for wavelengths λ where

$$\frac{2\pi d}{\lambda} \ll 1 \tag{1.10}$$

The pressure sensors 1 and 2 can be either in the side-by-side or face-to-face arrangement, as shown in figure 1.1. They are used to measure a single component of

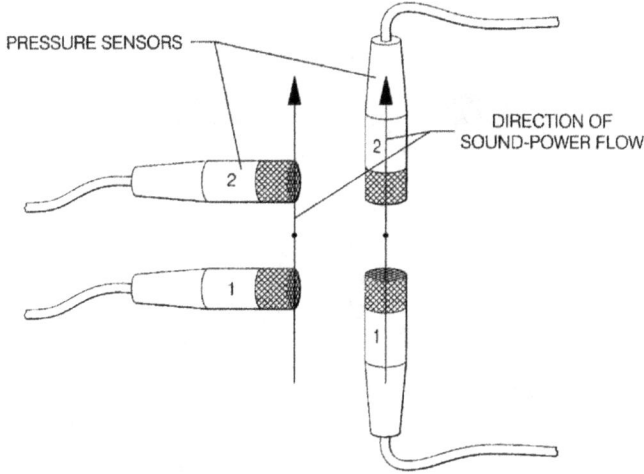

FIGURE 1

Figure 1.1. Arrangements of pressure sensors 1 and 2.

the sound-power flow vector in the direction r, where the spacing between sensors is $d = \Delta r$.

From equation (1.8), this component is

$$I_r = pv_r \tag{1.11}$$

where, from equation (1.2),

$$v_r = -\frac{1}{\rho_0 \Delta r} \int [p_2(t) - p_1(t)]\, dt \tag{1.12}$$

So that equation (1.11) becomes

$$I_r = \frac{p_2(t) + p_1(t)}{2\rho_0 \Delta r} \int [p_2(t) - p_1(t)]\, dt \tag{1.13}$$

The time-averaged sound-power flow, as a function of time, is then

$$I_r(t)_{\text{avg}} = -\frac{1}{2\rho_0 \Delta r} \cdot \frac{1}{t} \int_0^t \left\{ p_2(t) + p_1(t) \int [p_2(t) - p_1(t)]\, dt \right\} dt \tag{1.14}$$

Using Parseval's theorem, this expression was converted by Gary W Elko [4] to a function of frequency f

$$I_r(f)_{\text{avg}} = -\frac{1}{2\pi f \rho_0 \Delta r} \, \text{Im}[S_2(f) \cdot S_1^*(f)] \tag{1.15}$$

where $S_1(f)$ and $S_2(f)$ are the discrete Fourier transforms of the sound pressures recorded at the microphones 1 and 2. Im indicates the imaginary part of the cross spectrum $S_2(f) \cdot S_1^*(f)$, and the asterisk denotes the complex conjugate. Ways of calculating discrete Fourier transforms are discussed in texts, such as [5, 6].

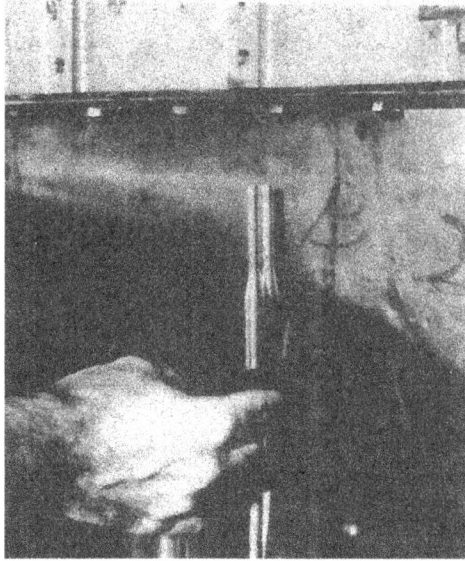

Figure 1.2. Measuring the sound-power flow at the surface of a diesel engine.

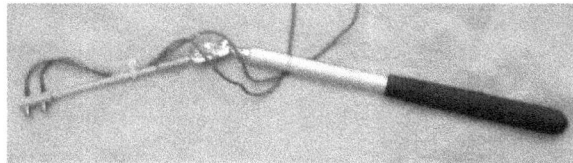

Figure 1.3. Probe with microphones in the side-by-side arrangement.

Equation (1.15) is the well-known cross-spectral relation for sound-power flow. Its validity has been demonstrated both theoretically [7–9] and experimentally [9–11]. Figure 1.2 illustrates a two-microphone probe in the side-by-side arrangement measuring the component of sound-power flow perpendicular to the surface of a diesel engine.

Figure 1.3 shows a version of a probe with two small sensitive microphones in the side-by-side arrangement.

The frequency range of measurement accuracy of sound-power flow depends on the microphone spacing d, as stated in equation (1.10), and is shown graphically in figure 1.4 for different values of d ranging from 5 to 50 mm.

1.3 Plane and spherical waves

The above formulae are simpler when the sound consists of harmonic waves of a single frequency f [1, 2]. Time dependence is expressed by $\exp(-\mathrm{i}2\pi f\,t)$, where $\mathrm{i} = \sqrt{-1}$. Equation (1.2) then becomes

$$\mathbf{v} = -\frac{\mathrm{i}}{\rho_0 2\pi f}\nabla p \qquad (1.16)$$

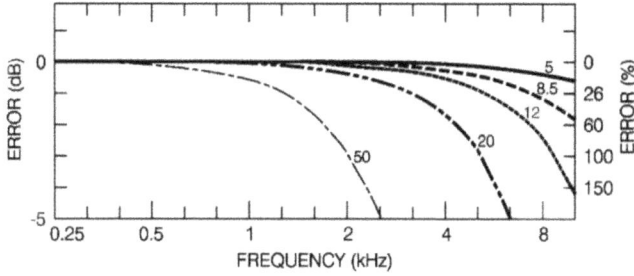

Figure 1.4. Measurement accuracy.

and the average sound-power flow per cycle is

$$I_{avg} = \frac{1}{2} \operatorname{Re}(v^*p) \tag{1.17}$$

where Re indicates the real part. Plane harmonic waves moving in the positive x direction can be expressed as $p = P \exp[ik(x - ct)]$, where $k = \frac{2\pi}{\lambda}$. Equation (1.17) then becomes

$$I_{avg} = \frac{P^2}{2\rho_0 c} \tag{1.18}$$

where the direction of the average sound-power flow is the direction of the plane waves.

Multi-frequency plane waves can be averaged over an extended period to give

$$I_{avg} = \frac{p^2_{rms}}{\rho_0 c} \tag{1.19}$$

where p_{rms} is the square root of the time-average of sound pressure squared and the sound-power flow is in the direction of the plane waves. Equation (1.19) is called the far-field approximation. In the past, this approximation has been used to measure the sound power of a source, by integrating it over an arbitrary imaginary surface enclosing the source [11]. The plane-wave assumption and the direction of the waves were frequently ignored. Pressure squared has mistakenly been called sound intensity in some branches of acoustics.

Harmonic spherical waves can be represented by

$$p = \frac{P \exp[-i(r - 2\pi f\, t)]}{r} \tag{1.20}$$

It can be shown [1, 2] that equation (1.20) satisfies equations (1.18) and (1.19).

1.4 Measuring the vector sound-power-flow with different types of probes

The three-dimensional vector for sound-power flow in fluids is determined using four omnidirectional pressure sensors at the vertices A, B, C, and D of a regular

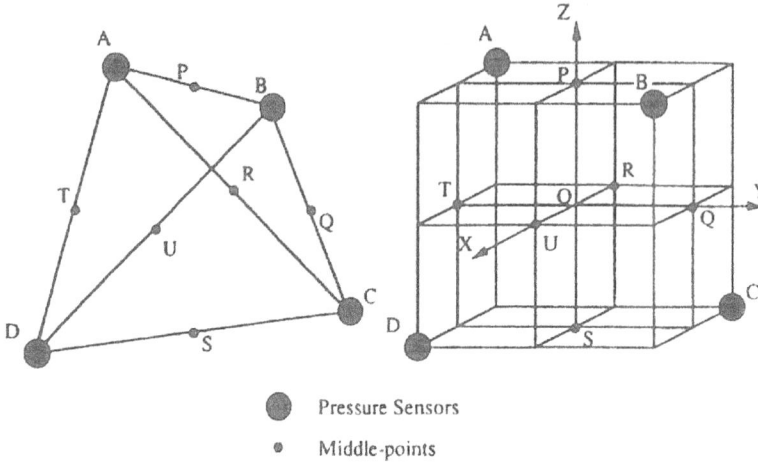

Pressure Sensors

Middle-points

Figure 1.5. Regular tetrahedral arrangement.

(a)   (b)

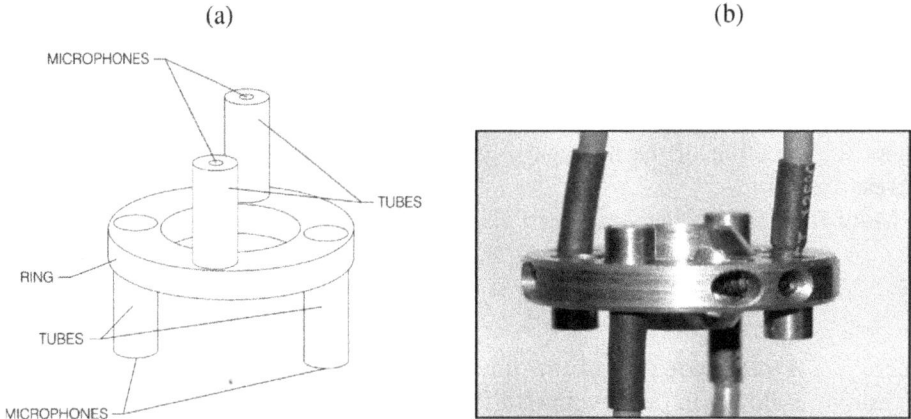

Figure 1.6. Full-space probe.

tetrahedron, as shown schematically in figure 1.5(a). A regular tetrahedron has edges of the same length. It has the property that the joins of midpoints of opposite edges form a Cartesian set of coordinates, as shown in figure 1.5(b). The components of the vector are then obtained using the cross-spectral formula in equation (1.15).

The sound pressures at the midpoints are determined using finite-difference approximations. The approximations are valid if the distance d, along a tetrahedral edge between any two sound-pressure sensors, satisfies the relation $d \ll \lambda/2\pi$, where λ is the wavelength of sound. The wavelength λ is equal to the speed of sound c in the medium, divided by f, the acoustic frequency. Hence the upper limit for d is given by

$$d \ll c/(2\pi f). \tag{1.21}$$

Figure 1.6 shows a probe for measuring the sound-power flow vector in three-dimensional space. The probe has four small omnidirectional microphones pointing

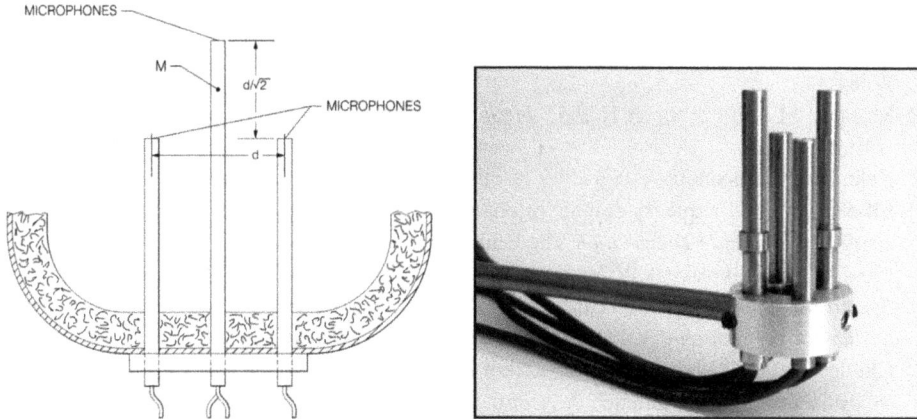

Figure 1.7. Half-space probe.

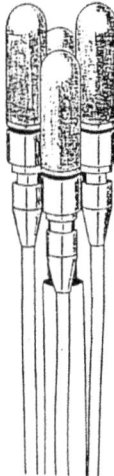

Figure 1.8. Probe for use in water.

in opposite directions at the ends of narrow tubes attached to a ring. This is called the full-space probe [12].

In applications where the probe is on the ground or attached to a wall, the sound-power flow-vector is measured in a half space. Figure 1.7 shows a half-space vector probe, with backing to prevent interference from reflections from behind the probe [13].

Figure 1.8 shows the regular tetrahedral arrangement of hydrophones for use in water. This can be used as a full-space probe, except in the direction immediately behind the probe.

References

[1] Kinsler L E, Frey A R, Coppens A B and Sanders J V 1982 *Fundamentals of Acoustics* 3rd edn ch 5 (New York: Wiley)

[2] Pierce A D 1997 Mathematical theory of wave propagation ch 2 **1** 21–38
Pierce A D 1997 *Encyclopedia of Acoustics* ed M J Crocker Editor-in Chief (New York: Wiley)

[3] Morse P M and Feshbach H 1953 *Methods of Theoretical Physics* (New York: McGraw-Hill) pp 141–3

[4] Reference 1, Appendix A10, Tables of Physical Properties of Matter

[5] Elko G W 1984 Frequency domain estimation of the complex acoustic intensity and acoustic energy density *PhD Dissertation* The Pennsylvania State University

[6] Press W H, Teukolsky S A, Vetterling W T and Flannery B P 1992 *Numerical Recipes in Spectral Applications* (Cambridge: Cambridge University Press) FFTW, A C subroutine library for computing discrete Fourier transforms http://www.fftw.org

[7] Chung J Y 1978 Cross-spectral method of measuring acoustic intensity without error caused by instrument phase mismatch *J. Acoust. Soc. Am.* **64** 1613–16

[8] Fahy F J 1995 *Sound Intensity* 2nd edn (London: Chapman and Hall) Chapters 4 & 5

[9] Chung J Y, Pope J and Feldmaier D A 1979 Application of acoustic intensity measurement to engine noise evaluation SAE paper 790502 *Proc. Diesel Engine Noise Conf.* P-80

[10] Pope J, Hickling R, Feldmaier D A and Blaser D A 1981 The Use of acoustic intensity scans for sound power measurement and for noise source identification in surface transportation vehicles SAE paper 810401

[11] ANSI STANDARD, Engineering methods for the determination of sound power levels of noise sources for essentially free-field conditions over a reflecting plane, ANSI S1.34-1980 (ASA-14-1980), (1980)

[12] Hickling R 2006 Acoustic measurement method and apparatus *US Patent No.* 7,058,184 B1

[13] Hickling R 2011 Vector sound-intensity probes operating in a half space *US Patent No.* 7,920,709 B1

Chapter 2

Calibration and normalization of probes in air

Calibration and normalization are performed using the instrumentation [1] shown below in figure 2.1. This consists of a driver or speaker at one end of a tube, which transmits plane waves incident on microphones held in a fixture at the other end of the tube. The microphones are those in the probes shown in the previous chapter in figures 1.3, 1.6, and 1.7, combined with a standard calibration microphone C. Standing waves in the tube are eliminated using banks of quarter-wave attenuators with openings flush with the inside wall of the tube. The standing waves are even and odd, as illustrated respectively in figures 2.2(a) and 2.2(b). Transfer-function methods are used to calibrate and normalize the microphones in a probe. The standard microphone C is calibrated first before being inserted in the fixture, using the acoustical characteristics provided by the manufacturer. Transfer-function methods are then used to normalize, i.e. to make the response of the microphones in a probe identical with C. These methods are described below.

The use of transfer functions in the normalization and calibration procedure is mathematically as follows. Standard digital Fourier transform (DFT) techniques are performed in a microprocessor to determine the transfer function H1C(f) between microphone 1 (for example) in a probe and the standard microphone C, as follows

$$H1C(f) = G1C(f)/G11(f) \qquad (2.1)$$

where G1C(f) is the cross-spectrum between the signal at microphone 1 and the calibration microphone C, given by

$$G1C(f) = F_pC(f) \cdot F_p1(f)^* \qquad (2.2)$$

and G11(f) is the auto-spectrum of the signal at microphone 1 given by

$$G11(f) = F_p1(f) \cdot F_p1(f)^* \qquad (2.3)$$

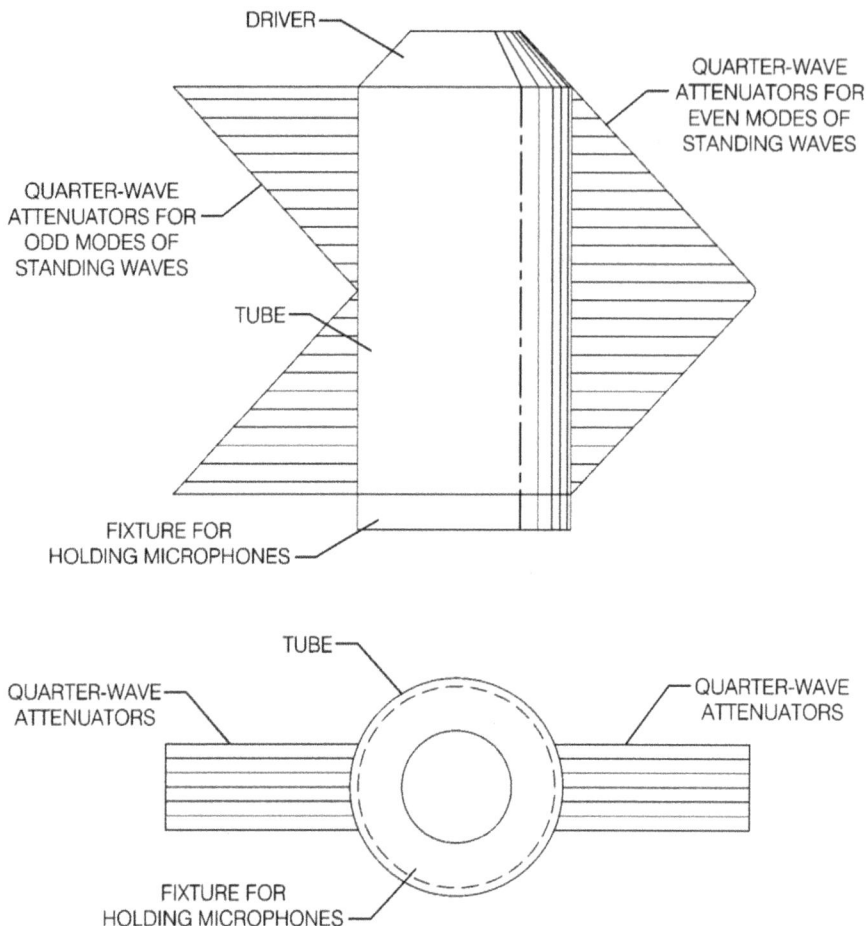

Figure 2.1. Calibration and normalization instrument.

where the asterisks denote the complex conjugate. To make the signal $Fp1(f)$ at microphone 1 in the probe look like the signal $FpC(f)$ at the calibration microphone C, it is multiplied by the transfer function in equation (2.1) to give

$$Fp1C(f) = Fp1(f) \cdot H1C(f) \qquad (2.4)$$

The process is repeated for microphone 2 using relations corresponding to equations (2.1) through (2.4) with 2 substituted for 1, as follows

$$H2C(f) = G2C(f)/G22(f) \qquad (2.5)$$

where

$$G2C(f) = FpC(f) \cdot Fp2(f)^* \qquad (2.6)$$

(a)   (b)

EVEN MODES   ODD MODES

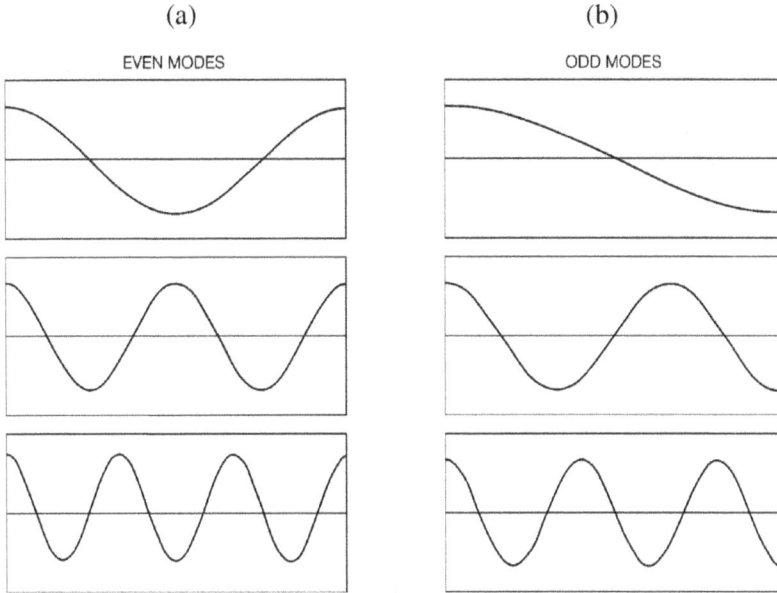

Figure 2.2. Standing waves in a tube.

Figure 2.3. Calibration and normalization instrument.

and

$$G22(f) = Fp2(f) \cdot Fp2(f)^* \qquad (2.7)$$

To make $Fp2(f)$ look like $FpC(f)$, $Fp2(f)$ is multiplied by the transfer function in equation (2.5) to give

$$Fp2C(f) = Fp2(f) \cdot H2C(f) \qquad (2.8)$$

The fixture in figure 2.4(a) is used for a two-microphone probe. The fixture in figure 2.4(b) is used to obtain the transfer functions for the microphones facing in one

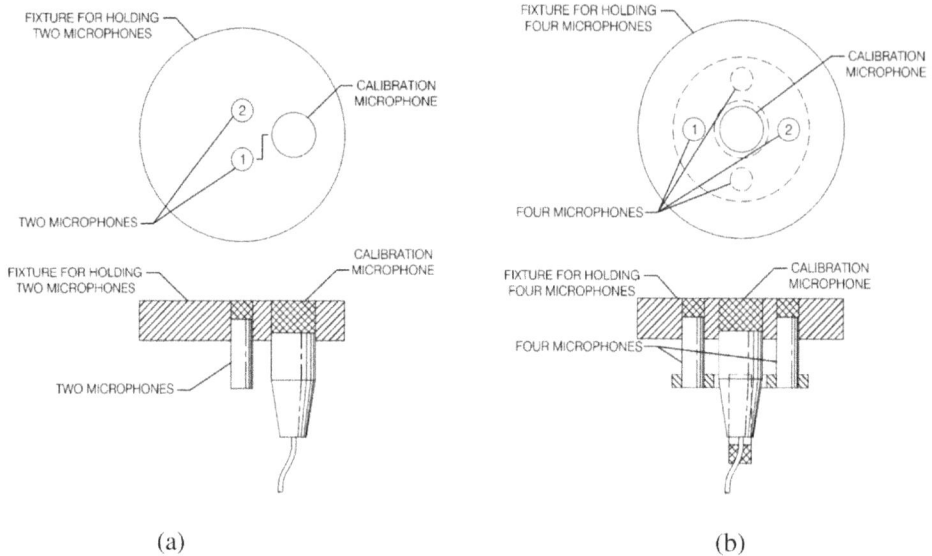

(a)   (b)

Figure 2.4. End fixtures for calibration equipment.

direction of the full-space probe in figure 1.6. The transfer functions for the microphones facing in the opposite direction are obtained in the same way by reversing the probe and inserting the standard microphone C in the fixture. In this way all four microphones in the probe can be made to look like microphone C, making the sensitivity of the probe omnidirectional and calibrating the individual microphones using the known acoustical characteristics of C. The transfer functions are stored in a digital signal processor for later use in measurements with the probes. Calibrations based on the known acoustical characteristics of C are applied in the digital signal processor for accuracy of measurement.

Reference

[1] Hickling R 2006 Normalization and calibration of microphones in sound-intensity probes, *US Patent No.* 7526094 B2

Chapter 3

Mathematics of sound-power flow in solids

3.1 Introduction

The theory of sound waves in solids requires knowledge of vectors, matrices, and tensors [1, 2]. Use is made of the theorem that a vector $\mathbf{F}$ is the sum of a gradient, and a curl

$$\mathbf{F} = \text{grad } \phi + \text{curl } \mathbf{A} \tag{3.1}$$

where grad $\phi = \nabla \phi$ and curl $\mathbf{A} = \nabla \times \mathbf{A}$. ϕ is called the scalar potential of $\mathbf{F}$. $\mathbf{A}$ is called the vector potential, where div $\mathbf{A} = \nabla \cdot \mathbf{A} = 0$. Also needed are the two elastic constants λ and μ, where λ is the Lamé constant and μ is the shear modulus. Their units are in Pascals. Other elastic constants can be derived from these. For example Young's modulus is $E = \frac{\mu(3\lambda + 2\mu)}{\lambda + \mu}$ and Poisson's ratio is $\sigma = \frac{\lambda}{2(\lambda + \mu)}$. An elastic constant can be derived from any two of the other constants. Values for different solids are given in appendix B, and in [1, 2], and other sources. For example, aluminum has $\rho_0 = 2700$ kg m^{-3}, $\mu = 25$ GPa, and $\lambda = 61$ GPa, and the longitudinal (compressional) velocity $c_l = 6418$ m s^{-1} and shear (transverse) velocity $c_s = 3046$ m s^{-1}. Transverse means perpendicular to the direction of the wave. The scalar potential ϕ for longitudinal (compressional) waves is obtained by solving the equation

$$\nabla^2 \phi = \left(\frac{1}{c_1}\right)^2 \frac{\partial^2 \phi}{\partial t^2} \tag{3.2}$$

where $c_1^2 = \frac{\lambda + 2\mu}{\rho_0}$. The vector potential $\mathbf{A}$ for transverse (shear) waves is obtained by solving the equation

$$\nabla^2 \mathbf{A} = \left(\frac{1}{c_s}\right)^2 \frac{\partial^2 \mathbf{A}}{\partial t^2} \tag{3.3}$$

doi:10.1088/978-1-6817-4453-7ch3

where $c_s^2 = \frac{\mu}{\rho_0}$. The elastic displacement vector $\mathbf{u}$ is then given [2] by

$$\mathbf{u} = -\nabla\phi + \nabla \times \mathbf{A} \tag{3.4}$$

3.2 Harmonic radial waves

Harmonic radial waves are treated as previously in equation (1.20) for fluids. It follows that the vibrational velocity vector $\mathbf{v}$ is related to the displacement vector $\mathbf{u}$ by

$$\mathbf{v} = -i \, 2\pi f \, \mathbf{u} \tag{3.5}$$

The average sound-power flow per cycle is then given by

$$I_{\text{avg}} = -\frac{1}{2} \operatorname{Re}(v * \tau) \tag{3.6}$$

where τ is the stress tensor

$$\tau = \lambda \, \mathfrak{J} \, \nabla \cdot \mathbf{u} + \mu G \tag{3.7}$$

where $\mathfrak{J}$ is the identity tensor and G is the strain tensor [2].

References

[1] Pierce A D 1997 Mathematical Theory of Wave Propagation ch 2, Vol 1, pp 21–38
    Pierce A D 1997 *Encyclopedia of Acoustics* ed M J Crocker (New York: Wiley)
[2] Morse P M and Feshbach H 1953 *Methods of Theoretical Physics* (New York: McGraw-Hill) pp 141–3

Chapter 4

Sound sources and methods of exciting sound

4.1 Introduction

Sound sources can be divided into two types, natural and man-made. Examples of natural sources are: animals, wind, flowing streams, avalanches, and volcanoes. Examples of man-made sources are: airplanes, helicopters, road vehicles, trains, explosions, factories, and home appliances such as vacuum cleaners and fans. Sound excitation can be divided into two areas: aero (or hydro)-acoustics and structural acoustics. In the first, sound is generated by fluid motion. In the second, sound is generated by structural vibration. These can occur in combination. Sound can either be air-borne, structure-borne, or both. A few more significant examples of sound excitation are described here.

4.2 Engine knock

An example of sound excitation is the noise caused by cavity resonances in the combustion chambers of automotive engines [1]. In gasoline engines, this is called knocking and is due to auto-ignition of the end gas during combustion. Knock detectors are generally tuned to the lowest frequency of the cavity resonances. The characteristic knocking of diesel engines is due to the rapid pressure rise in compression ignition. Detailed information about engine noise is given in [2], which discusses excitation sources, transmission paths, structural vibrations, and noise radiation.

4.3 Turbulent flow

Another example of sound excitation is the noise due to turbulent flow in a fluid.

Pioneering work was performed by Lighthill [3], particularly for jet-engine noise. Turbulent flow is due to the viscosity of the fluid. If there is no viscosity, the fluid passes smoothly over a surface, as in figure 4.1.

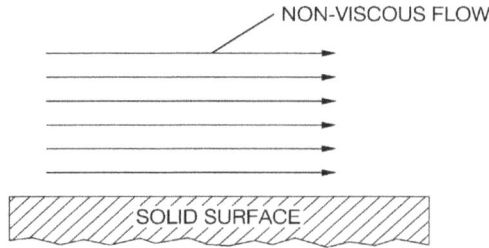

Figure 4.1. Non-viscous flow over a solid surface.

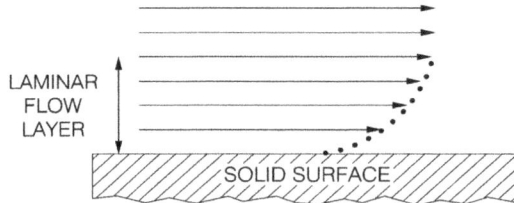

Figure 4.2. Laminar flow over a solid surface.

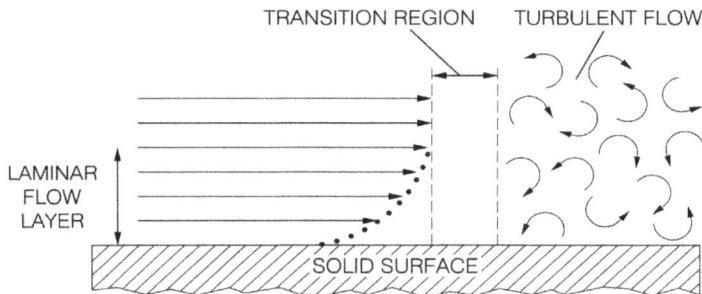

Figure 4.3. Transition to turbulent flow.

However, since fluids are viscous, viscosity causes the flow to adhere to the surface. The speed of the flow gradually increases away from the surface to form a laminar-flow layer, as shown in figure 4.2.

Generally the laminar-flow layer does not persist over the surface. It passes through a transition zone to form turbulent flow, as shown in figure 4.3.

Turbulent flow increases the drag of a body. Also noise is generated by turbulent flow over a surface, which becomes louder with increasing flow speed. Fish and sea mammals are able to delay the onset of turbulence in order to reduce both drag and noise. When the flow is not over a surface, there is no detectable noise. For example, figure 4.4 shows laminar flow changing into noiseless turbulence in cigarette smoke.

4.4 Cavitation and cavitation-erosion in liquids

Another example of a noise source is cavitation in liquids. Cavitation is the formation of bubbles in a liquid due to a local decrease in pressure. It is similar to boiling, where bubbles are formed by an increase in temperature. Cavitation can

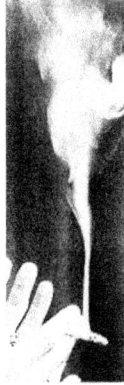

Figure 4.4. Noiseless turbulent flow.

occur when a liquid flows rapidly over a solid surface, such as on the suction side of a ship's rotating propeller. The collapse of bubbles in a cavitation cloud can create intense localized forces that cause erosion of the metal of the propeller. Lord Rayleigh [4] explained the erosion, using a spherical bubble in an incompressible liquid as a model. Collapsing cavitation bubbles also generate local shock waves and sound. This was demonstrated using a mathematical model of a bubble collapsing in a compressible liquid [4–6], as shown in figure 4.5, and later verified experimentally [7]. The numbers on the bubble wall in figure 4.5 indicate time progression.

In addition to compressing the gas inside the collapsing bubble, the liquid surrounding the bubble is highly compressed making it highly viscous. Momentarily it can then become similar to a solid, see chapter 3, supporting shear waves as well as compressional waves. As yet, this has not been investigated. Perhaps it could be investigated using single bubble luminescence described below.

A reentrant jet can form inside a collpsing bubble next to a surface, as shown in figure 4.6.

Cavitation noise is important for sonar detection of moving underwater objects, for example cavitation on the propellers of a submarine, as shown in figure 4.7. Fish and sea mammals appear to be able to avoid generating noticeable flow noise from cavitation.

To reduce noise and erosion, ship propellers are shaped differently from aircraft propellers, as shown in figure 4.8, where the ship propeller is shown on the left and the aircraft propeller is on the right.

The amount of cavitation erosion depends on the temperature of the water [8] and on the Brinell hardness (B/N) of the metal being eroded, as shown by the data in figure 4.9. Brinell hardness can be converted to mega-Pascals by multiplying by the acceleration due to gravity, 9.81 m s^{-2}.

4.5 Supercavitating structures

In order to increase the speed and range of underwater structures, cavitation can be induced at the leading edge of the structure with gas injected into the resulting cavity, as shown in figure 4.10.

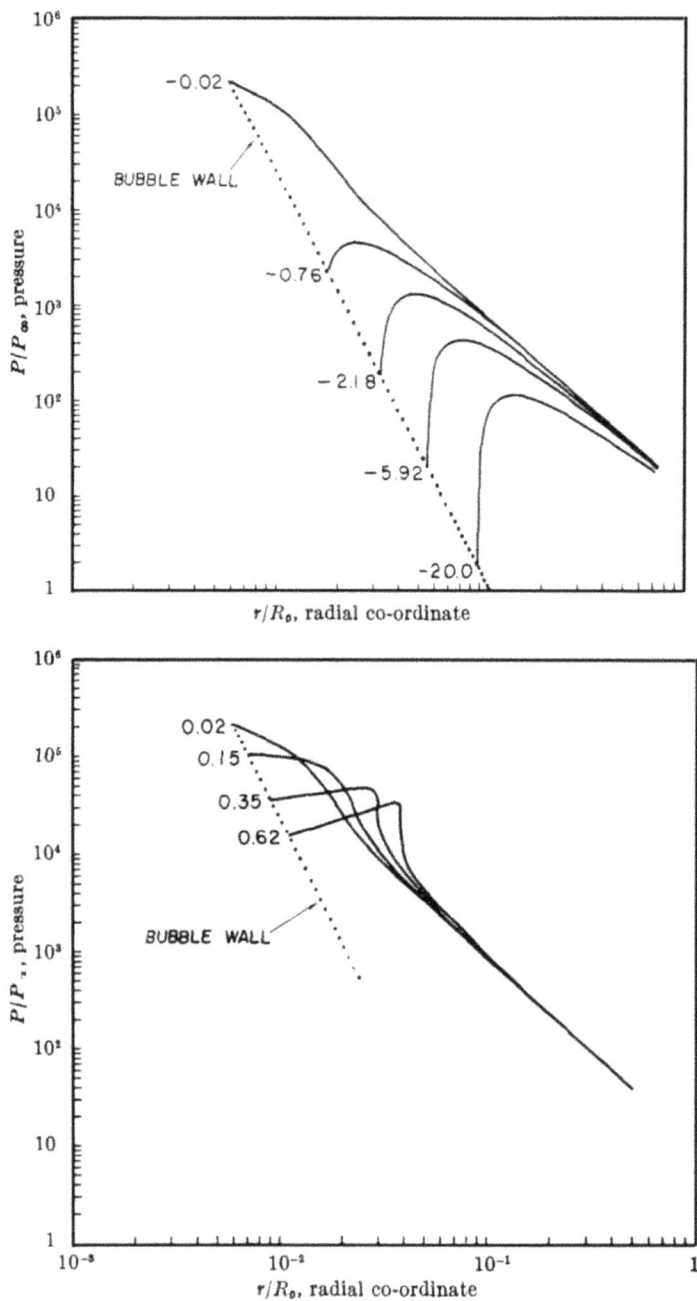

Figure 4.5. Collapse and rebound of a sperical bubble in water. (a) Collapse showing high compression of gas in a bubble. (b) Rebound showing formation of an outgoing shock wave.

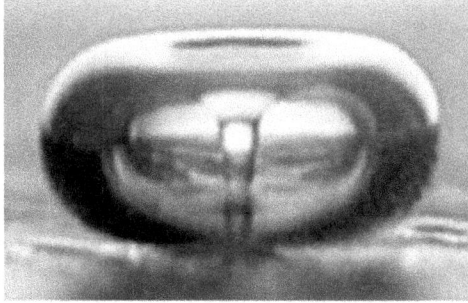

Figure 4.6. Reentrant jet in a cavitation bubble.

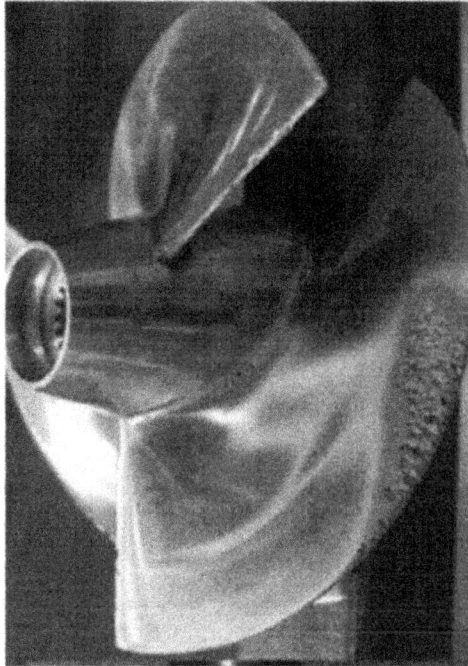

Figure 4.7. Cavitation on a propeller.

A supercavitating torpedo is shown in figure 4.11. In practice, it is difficult to steer a supercavitating vessel. A rudder won't work without water contact. Also developing a torpedo engine that operates over long distances is difficult.

4.6 Luminescence from cavity collapse

Another effect of cavity collapse is the luminescence that occurs when the gas inside a bubble is rapidly compressed to high temperatures. This effect is known as sonoluminescence. It occurs because there is not enough time [5, 6] during the collapse for the heat of the compressed gas in the bubble to be conducted into the surrounding liquid. The intensity of the luminescence is related to the thermal

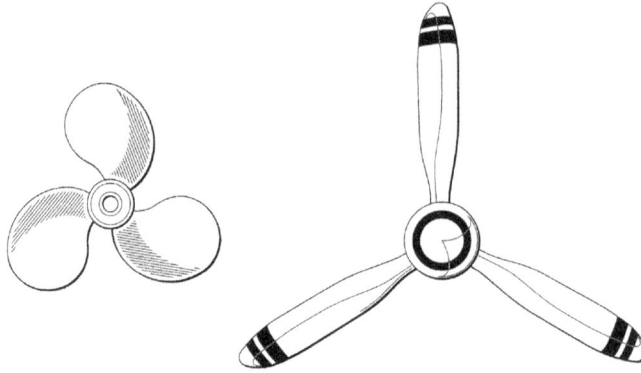

Figure 4.8. Ship propeller compared to an aircraft propeller.

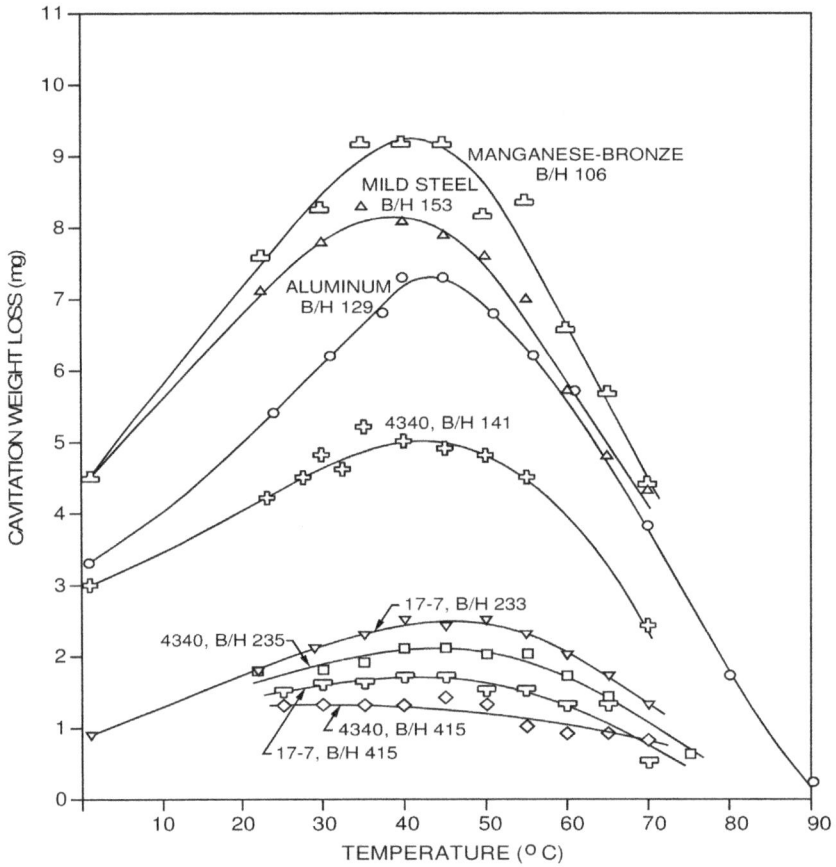

Figure 4.9. Effect of water temperature on cavitation erosion for different metals.

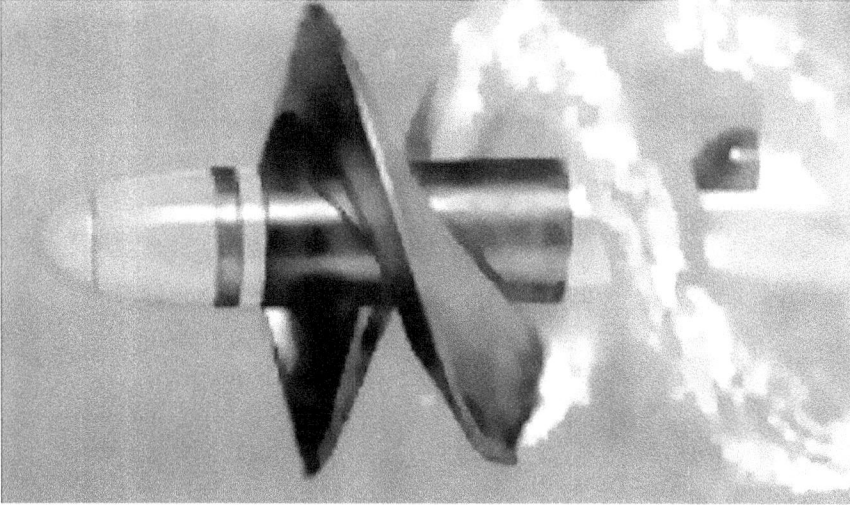

Figure 4.10. Supercavitating propeller.

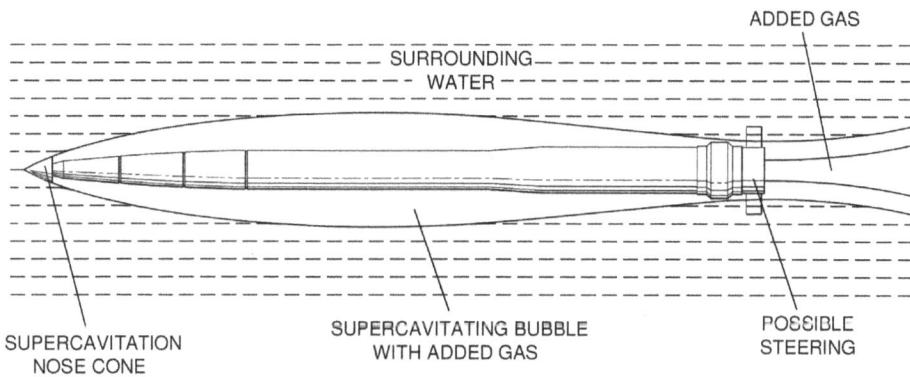

Figure 4.11. Supercavitating torpedo.

conductivity of the gas dissolved in the liquid [9]. The effect of cavity collapse is illustrated in figure 4.12, where the dashed lines show the rapid adiabatic compression of water at the bubble wall at two initial water temperatures. It is seen that compression at the bubble wall drives the liquid into the ice zones of water. However there is limited time for ice to form [10]. The figure also indicates why ice floats at atmospheric pressure. It has been postulated that sonoluminescence is caused by a converging shock wave in the gas inside the spherical bubble. However this cannot be correct. A moving interface does not generate a pressure wave propagating ahead of it. The gas pressure inside a collapsing spherical bubble is therefore uniform.

There are two forms of sonluminescence. One is called multi-bubble luminescence that occurs within the high-speed cavitation flow described above and in other ways. Figure 4.13 shows a photograph of multi-bubble luminescence, as seen through a light intensifier.

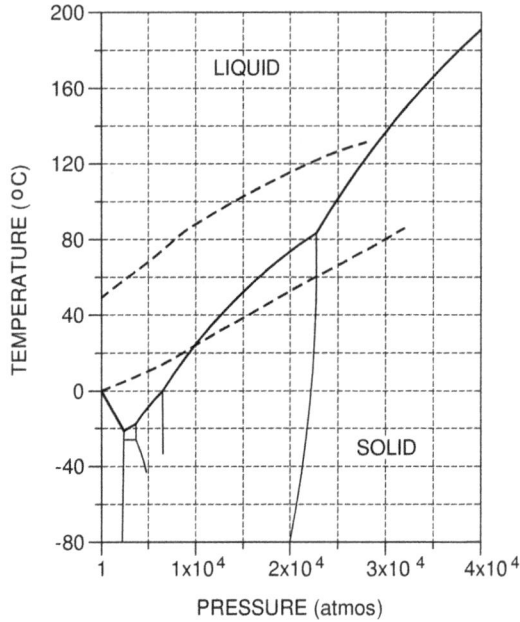

Figure 4.12. Rapid adiabatic compression in water (dashed lines).

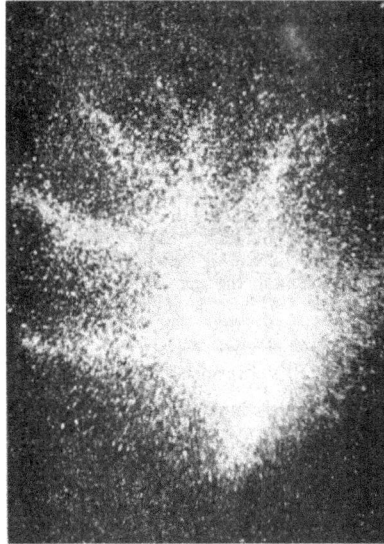

Figure 4.13. Multi-bubble luminescence.

The intensity of luminescence has been shown [9] to vary for solutions of different dissolved gases in water, as in table 4.1.

The other type of sonoluminesce is called single-bubble luminescence [11, 12], which can be formed using the driven-flask arrangement shown in figure 4.14, derived from a description in [13]. The flask is driven at resonance with its maximum

Table 4.1. Luminescence intensity for solutions of different dissolved gases in water.

Gas	Relative intensity of sonoluminescence	Thermal conductivity (Cal/cm/Sec/°C) $\times 10^3$
Helium	1	0.337
Neon	18	0.109
Argon	54	0.038
Krypton	180	0.021
Xenon	540	0.012
Hydrogen	0	0.414
Oxygen	35	0.058
Nitrogen	45	0.055

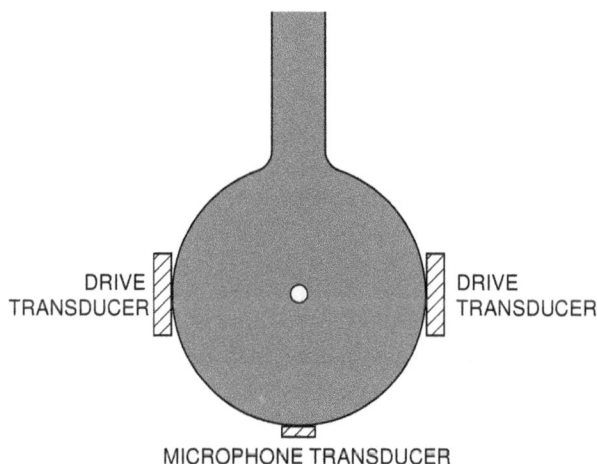

Figure 4.14. Single-bubble luminescence ocurring at the center of a driven flask.

amplitude at the center of the flask. This creates a single bubble that grows and collapses, giving the impresion of a single continuous light source.

Single-bubble luminescence has stimulated a lot of of interest, which was directed, to some extent, towards the possibility of generating nuclear fusion. So far there has been no success with this, although hope still remains. There has been much more success in the field of multi-bubble sonochemistry [14, 17]. This originited in the work of Virginia Griffing and coworkers [16], who understood the nature of cavity collapse and its chemical effect at an early time, as they indicate in the following statement:

'From the results it was concluded that primary reactions seem to be gas phase reactions, probably of a thermal nature, taking place inside the gas bubbles which serve as "hot spots" in the liquid. The calculations show that, under the experimental conditions reported here, temperatures of several hundreds or thousands [of] degrees can be easily reached inside the cavitating bubbles in resonance with the sound field;

it is believed that at these temperatures the reaction, which gives rise to H_2O_2 takes place'.

4.7 Cavity collapse and the origins of life

There are many explanations of the origins of life [17]. The mechanical and chemical effects of cavity collapse [14, 15], outlined in the foregoing, indicate how cavitation bubbles occuring in crashing waves, currents, and volcanic action in ancient oceans and rivers can create new molecules. Over time the new molecules can originate life.

4.8 Text describing the behavior of bubbles in liquids

A predecessor [18] of this book provides an extensive review of information on bubbles in liquids. However it does not include much of the material for practioners and research scientists given here.

References

[1] Hickling R, Feldmaier D A, Chen F H K and Morel J S 1983 Cavity resonances in engine combustion chambers and some applications *J. Acoust. Soc. Amer.* **73** 1170–78

[2] Hickling R and Kamal M M (ed) 1982 *Engine Noise: Excitation, Vibration and Radiation* (New York: Plenum)

[3] Lighthill M J 1961 Jet noise *AIAA Journal* **1** 1507–17

[4] Lord Rayleigh 1917 On the pressure developed in a liquid during the collapse of spherical cavity *Phil. Mag.* **34** 94–8

[5] Hickling R 1962 I. Acoustic radiation and reflection from spheres, II. Some effects of thermal conduction and compressibility in the collapse of a spherical bubble in a liquid *PhD Thesis*, California Institute of Technology, Pasadena, CA

[6] Hickling R and Plesset M S 1964 The collapse and rebound of a spherical bubble in water *Phys. Fluids* **7** pp 7–14

[7] Lauterborn W 1997 Cavitation ch 25, Vol. 1 pp 263–70
Lauterborn W 1997 *Encyclopedia of Acoustics* ed M J Crocker (New York: Wiley)

[8] Plesset M S 1962 Private communication

[9] Hickling R 1963 Effects of thermal conduction of dissolved gas in sonoluminescence *J. Acoust. Soc. Amer.* **35**

[10] Hickling R 1994 Transient high-pressure solidification associated with cavitation in water *Phys. Rev. Letters* **73** 2853–6

[11] Gaitan D F, Crum L A, Church C C and Roy R A 1906 Sonoluminescence and bubble dynamics for a single, stable, cavitation bubble *J. Acoust Soc. Amer.* **91** 3166–83

[12] Gaitan D F and Holt R G 1999 Experimental Observations of Bubble Response and Light Intensity near the Threshold for Single Bubble Sonoluminescence in an Air-Water System *Phys. Rev. E* **59**

[13] Barber B P and Putterman S J 1991 Observation of Synchronous Pico-second Sonoluminescence *Nature* **352** 318–20

[14] Suslick K S 1990 Sonochemistry *Science* **247** 1439–45

[15] Brennen C E 2001 Fission of collapsing cavitation bubbles *Fourth International Symposium on Cavitation*, California Institute of Technology.

[16] Fitzgerald M E, Griffing V and Sullivan J 1956 Chemical Effects of Ultrasonics, 'Hot Spot' Chemistry *J. Chem. Phys.* **25** 926

[17] Horowitz N H and Hubbard J S 1974 *The origin of life annual reviews*, www.annualreviews.org/aronline

[18] Leighton T G 1994 *The Acoustic Bubble* (London: Academic Press)

Chapter 5

Sound-power flow over the ground, and in the atmosphere and ocean

5.1 Sound-power flow over the ground

The travel of sound is greatly affected by passing over the ground [1]. Figure 5.1 shows a test of sound-power flow over the ground performed by the author.

Sound-power flow at ground level is affected by noise barriers, trees, and turbulence, particularly in urban environments [1] like cities where there are a series of tall buildings, such as in New York.

Figure 5.1. Test of sound-power flow over the ground.

5.2 Sound-power flow in the atmosphere and ocean

It is known that the transmission of sound in the atmosphere is affected by wind and temperature. Detailed information about sound transmitted in the atmosphere is provided in the literature [2]. Temperature generally decreases with altitude, causing an upward refraction of sound. This can create a dead zone of transmitted sound at a distance from the source. On the other hand, wind speed generally increases with altitude creating a downward refraction. A combination of the two can cause sound to be heard in a zone some distance from the source, but not between.

In the ocean, an important feature of sound propagation is the Sound Fixing and Ranging (SOFAR) channel. Generally temperature in the ocean decreases with depth causing a downward refraction, while pressure increases with depth causing an upward refraction. The combination traps sound in the SOFAR channel, within which the sound can travel over long distances [3].

5.3 Obstruction of sound-power flow by particles in a fluid

Backscattering by particles suspended in a fluid can obstruct sound-power flow. The frequency of the sound determines the amount of obstruction and backscattering. Figure 5.2 shows the backscattering amplitude calculated [4] as a function of the non-dimensional frequency ka, where $k = 2\pi f/c$ and a is the particle radius, c being the speed of sound in the fluid. In the figure, ξ indicates particle density divided by fluid density.

Figure 5.2 shows that the backscattered amplitude is small when ka is small, in particular that there is very little scattering when ka is less than 0.1, i.e. when the frequency satisfies the relation

$$f < 0.1c/2\pi a \tag{5.1}$$

Fog particles in air and sediment particles in water provide examples of scattering. Fog particles range from 1 to 14 μm in diameter [5] and the speed of sound c in air is about 345 m s^{-1}. Audible sound at frequencies below about 20 kHz penetrates fog.

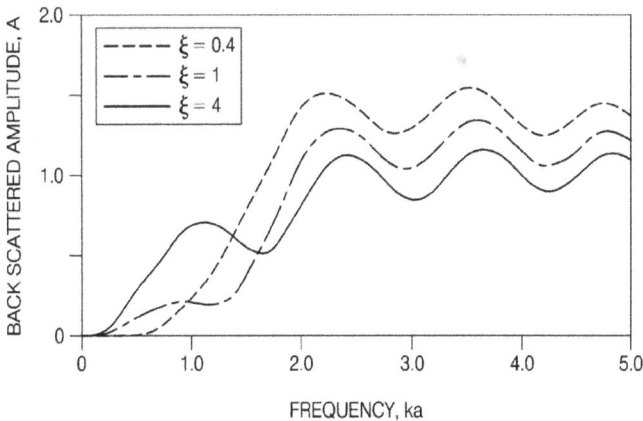

Figure 5.2. Backscattering by particles in a fluid.

For example, foghorns demonstrate how low-frequency sound can carry over long distances. A similar effect occurs with motor vehicle and train lights in a fog. Normal white light is scattered by fog. Greater penetration of fog can be achieved using red light, which is the longest visible wavelength. However use of red has to be avoided, because it indicates danger. For this reason, orange, the next longest wavelength is used.

Sediment particles in water have various constituents, such as sand, silt (mostly rock particles) and clay [6]. The size of the particles can vary over a wide range. The particle size of medium-sized sand ranges from 1.25–0.25 mm. Finer sand particles have a diameter less than 0.075 mm, as determined by passing through a #200 sieve. The speed of sound in salt water is about 1500 m s^{-1}. For fine sand, equation (5.1) shows that sound can penetrate sediment-laden water when the frequency is less than about 60 kHz. For larger particles and agglomerates with a diameter of roughly 0.25 mm, equation (5.1) indicates that the frequency has to be less than about 20 kHz for adequate penetration. It is well known that bubbles in water scatter sound quite strongly and are often used to obscure submarines and other underwater objects. Also bubbles can adhere to sediment particles, enabling the particles to float, increasing their ability to scatter sound.

References

[1] Attenborough K, Li K M and Horoshenkov K 2007 *Predicting Outdoor Sound* (London: Taylor and Francis)

[2] Everest F 2001 *The Master Handbook of Acoustics* (New York: McGraw-Hill) pp 262–3

[3] Munk W, Worcester P and Wunsch C 1995 *Ocean Acoustic Tomography* (Cambridge: Cambridge University Press)

[4] Hickling R and Wang N M 1966 Scattering of sound by a rigid movable sphere *J. Acoust. Soc. Am.* **39** 276–9

[5] Arnott W P *et al* 1997 Droplet size spectra and water-vapor concentration of laboratory water clouds *Appl. Opt.* **36** 5205–16

[6] Johnson L J 1979 *Introductory Soil Science: A Study Guide and Laboratory Manual* (New York: MacMillan)

Chapter 6

Sound-power flow in the Earth's structure

6.1 Model of the Earth's structure

Since the Earth is closely spherical in nature, its structure can be modeled using spherical co-ordinates [1]. The Earth's structure is shown below in figure 6.1. Our direct knowledge is confined to the outer crust, namely the oceans, continents, mountains, and the atmosphere. Knowledge of the interior of the crust is known from mines, drilling for oil, and water exploration. Fracking and oil drilling in the outer crust are illustrated in figure 6.2. Compared to the rest of the Earth's structure,

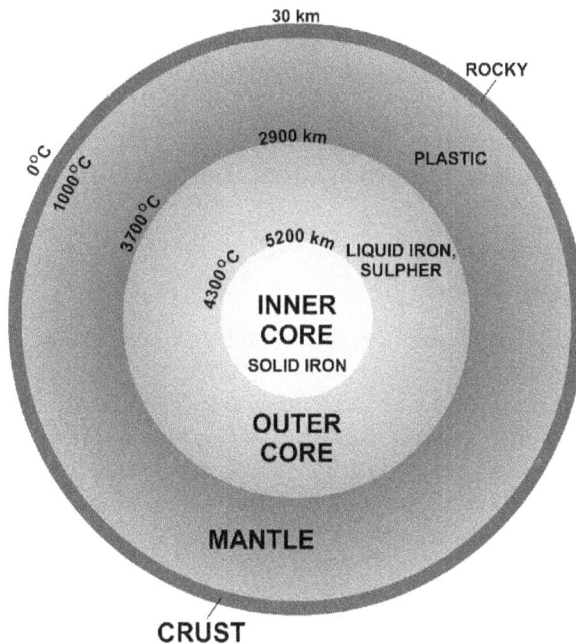

Figure 6.1. Model of the cross-section of Earth.

doi:10.1088/978-1-6817-4453-7ch6

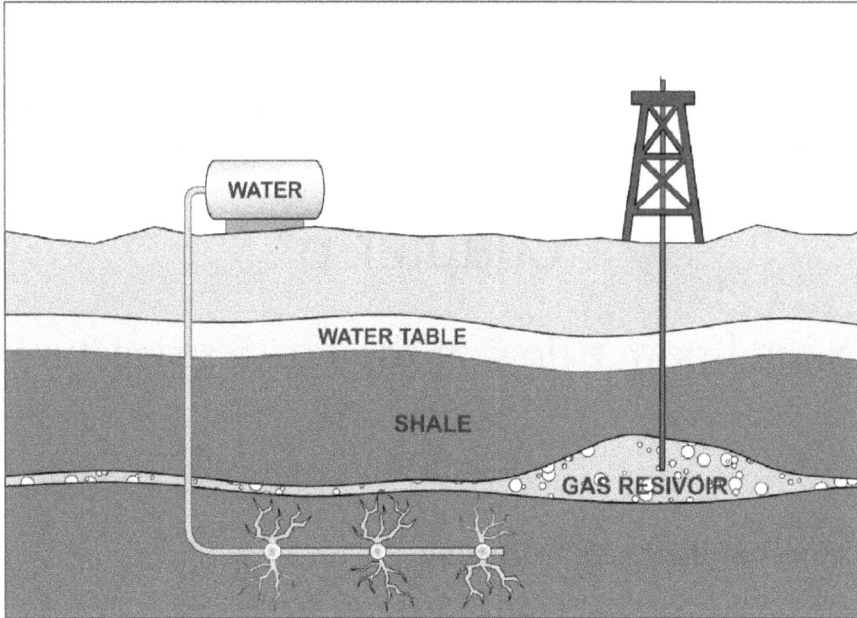

Figure 6.2. Model of fracking, and oil and water drilling.

the crust is brittle and can fracture in the form of earthquakes and volcanic eruption [2]. The Earth's interior below the outer crust is divided sequentially into mantle, outer core and inner core, which are subject to increasing pressure towards the Earth's center. Most of the Earth's mass is in the mantle. Our knowledge of the Earth, below the outer crust, has been determined largely from seismological data. The mathematics of sound-power flow in fluids and solids is given in chapters 1 and 3 above.

Oil is formed when large quantities of dead organisms, such as zooplankton and algae, are buried beneath sedimentary rock and subjected to intense heat and pressure.

6.2 The Earth's magnetic field

Because magnetic materials lose their magnetism when they are hotter than about 750 K, the Earth cannot be a large permanent magnet. Almost all of the Earth is hotter and the only other way to make a magnetic field is with a circulating electric current. Circulation and convection of the molten iron in the outer core create the Earth's magnetic field. This is driven by the release of energy as the iron oscillates from solid to liquid at the interface with solid iron of the inner core. This oscillation is unstable and over time (about 20 000 years), it can cause a change in direction of the magnetic field.

6.3 Wobble of the Earth's axis

Irregularities in the Earth's motion are caused by geophysical processes, such as the momentum exchange between ocean currents and the solid Earth and the exchange of mass between polar ice sheets and ocean. These have been shown to cause a wobble of the Earth's axis [3].

6.4 Tides

Tides in the Earth's seas and oceans are caused by the attraction of the Moon and sun as they move around the Earth.

References

[1] Hickling R 1985 *Lecture presented in Geology Department* (California Institute of Technology)

[2] Lay T and Wallace T C 1995 *Modern Global Seismology* (London: Academic)

[3] Munk W 2015 http:/--www.nytimes.com-2015-08-25-science-walter-munk-einstein-of-the-oceans-at-97.html%3f_r=0

Chapter 7

The sound power of a source measured on a hypothetical enclosing surface

7.1 Mathematical introduction

The sound power of a source measures the total sound-power flow in watts from a source. Figure 7.1 shows a cross section of an arbitrary hypothetical surface S enclosing a source X, where I_n is the component of the sound-power-flow vector perpendicular to S.

Using Gauss's theorem, the sound power emitted by source X is the integral of I_n over S, represented mathematically [1] by

$$\text{Sound power} = \int_S I_n \, dS \tag{7.1}$$

I_n is measured with a two-microphone probe. Sources outside S do not contribute to the sound power. The surface S can consist of an imaginary hemisphere set on a rigid base plane, as in figure 7.2. Since I_n is zero on the rigid base plane, the integration is performed only on the hemisphere.

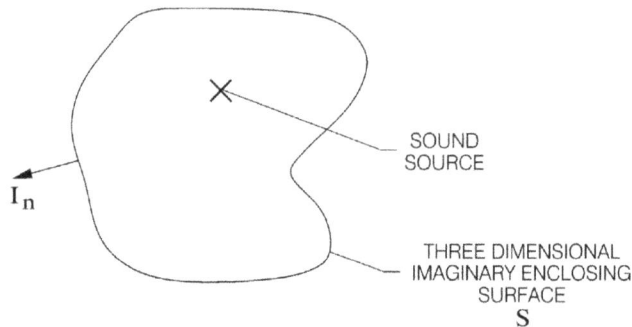

Figure 7.1. Measuring the sound power of a source.

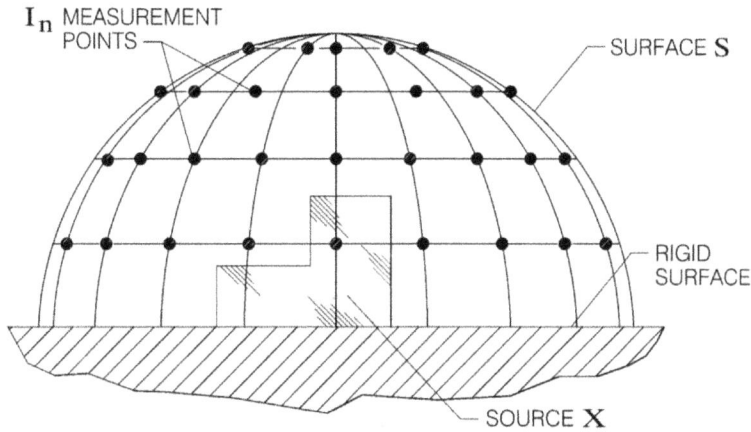

Figure 7.2. Measuring sound power on a rigid plane.

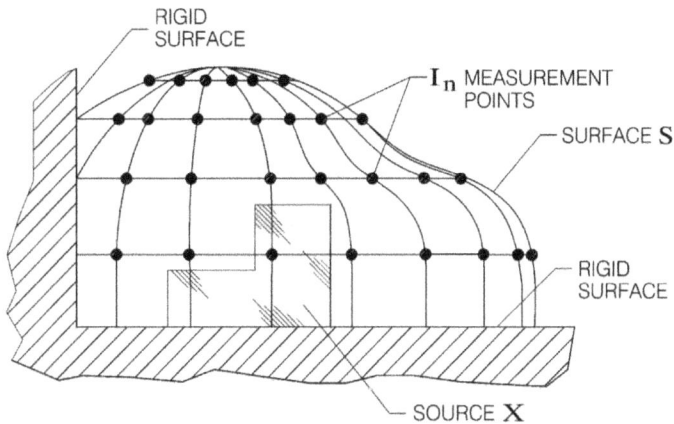

Figure 7.3. Measuring sound power with adjacent rigid surfaces.

Other rigid surfaces can form part of S, as shown in figure 7.3.

A measurement hemisphere on a rigid base plane is convenient, because an array of two-microphone probes on a semicircle can be rotated in equal steps over the hemisphere to obtain the sound power. Integration over the hemisphere is easiest if each measurement I_n is associated with an equal element of area. Also integration is more accurate if measurements of I_n are made at the center of each equal element of area. The integration process then consists of two steps. First the sound-power flow measurements are summed. The sum is then multiplied by the equal element of area to give the sound power of the source in watts.

The equal elements of area and the positions of the measurement points at the center of each element are determined using the well-known theorem of Archimedes illustrated in figure 7.4. The theorem states that an element of area on the surface of a hemisphere radius R has the same area as the element projected on to a cylinder enclosing the hemisphere.

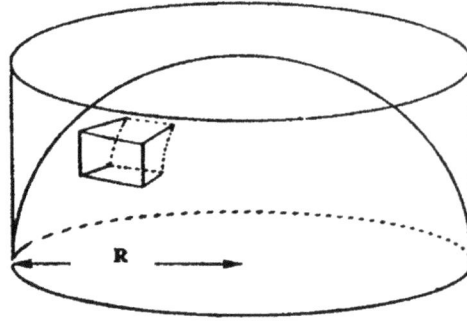

Figure 7.4. Archimedes' theorem.

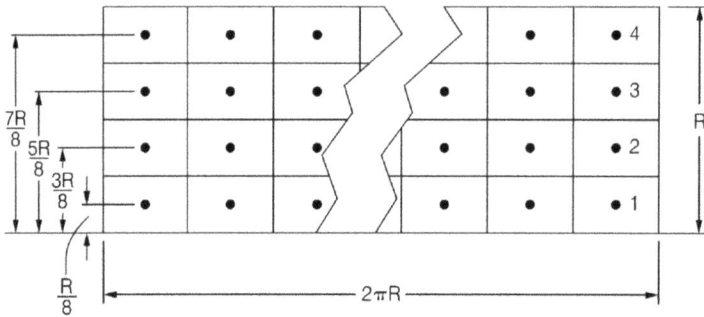

Figure 7.5. Unwrapped cylinder using Archimedes' theorem.

Figure 7.5 is the unwrapped cylinder showing the equal elements of area for eight two-microphone probes on a semicircular array.

The figure shows the height of the center of each element above the base plane. This in turn determines the height of each two-microphone probe in the semicircular array. The accuracy of the sound-power measurement can be checked using a standard sound-power reference source supplied by acoustical-instrument manufacturers.

7.2 Sound-power measurement in different test areas

The sound power of a source has been measured in different test areas. This was demonstrated [2] in tests with the semicircular array of two-microphone probes shown in figure 7.6. The array has a radius of 2 m. The source was a steady state, standard sound-power reference source, Bruel & Kjaer Model (B&K) 4204. The semicircular array is rotated in equal steps over an imaginary hemisphere set on a concrete floor. Concrete floors are not completely rigid. But if they are strongly built with steel rods and with a sealed surface, the rigidity assumption is generally valid.

The sound-power reference source has a manufacturer's calibrated value of 3.46 ± 0.25 mW. Results for the different test areas are given in table 7.1.

Figure 7.6. Measuring sound power in a reverberation room.

Table 7.1. Sound-power results for different test areas.

Test Space	Number Of Tests	Mean Sound Power (mW)	Standard Deviation
Semi-Anech. Room	6	3.37	0.02
Open Work Space (4m)	6	3.34	0.02
Open Work Space (2m)	6	3.4	0.02
Reverb. Room A	6	3.3	0.05
Reverb. Room B	3	3.36	0.01
Reverb. Room C	3	3.39	0.01

Of particular interest are the tests in three reverberation rooms. In a reverberation room sound is reflected back and forth, forming what is known as a diffuse sound field. In a diffuse field, sound pressure is believed by some to be uniformly distributed. This idealized concept implies that sound-power flow is zero in a reverberation room. However, sound power has to flow into the room from a speaker or other source, and out through the walls, ceiling and floor. Hence sound-power flow cannot be zero. Moreover, the accuracy of the results in table 7.1 for three very different reverberation rooms indicates that sound-power flow can be measured accurately in a reverberation room. Reverberation rooms A and B are rectangular, with dimensions respectively $6.7 \times 5.5 \times 4.1$ m and $8.8 \times 6.9 \times 5.5$ m. In an effort to make the sound field more diffuse, room C has non-parallel walls with a volume roughly 200 m^3. In the open workspace, in addition to the 2 m array, a larger semicircular array of 4 m was used.

The sound power of the B&K standard reference source was determined for the manufacturer in a semi-anechoic room, as illustrated in figure 7.7.

The measured sound power in table 7.1 is consistently slightly less than the manufacturer's value.

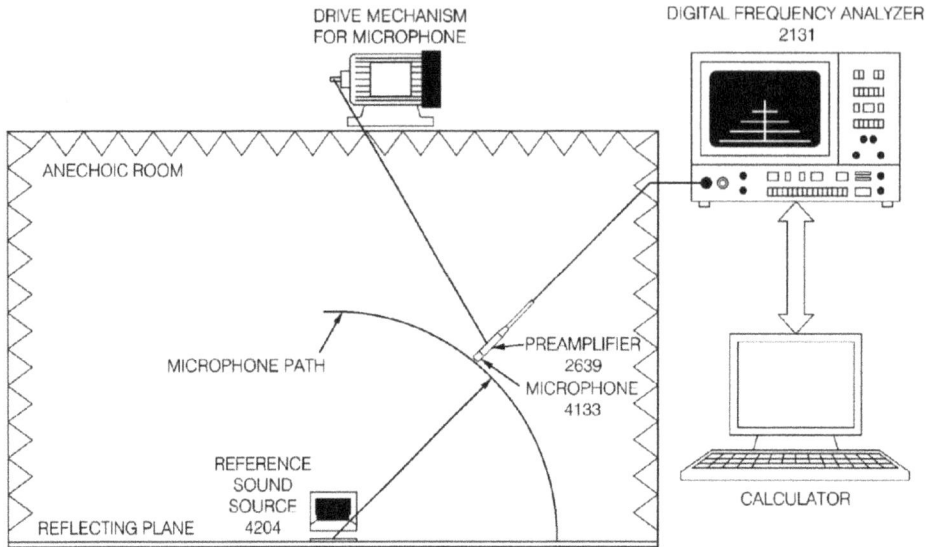

Figure 7.7. Measuring the sound power of a standard reference source.

7.3 Measuring the sound power of light vehicles

Indoor tests were conducted some years ago [3] with six gasoline-powered light vehicles. An indoor sound-power test can be used to replace existing outdoor passby tests, thus avoiding delays due to bad weather. In addition, the sound-power test can be used as a diagnostic tool for identifying faulty components.

The indoor sound-power test is illustrated in figure 7.8. Airflow from a fan external to the enclosing surface was used to cool the engine. The accuracy of the test was verified using a standard reference source. As shown in the figure, a semicircular array of two-microphone probes is rotated around a vehicle on a dynamometer roll set in the concrete floor. Sound power was determined for a range of fixed speeds and loads. The exhaust was allowed to flow freely into the test area and vented from the room every few steps in the rotation of the array. The noise of the dynamometer roll and the tires were reduced to low levels using the isolation system shown in figure 7.9 and their sound power was measured with the vehicle in neutral. This was subtracted from the sound power with the vehicle in drive. In this way it is possible to measure the sound power of the intake, power train and exhaust separately from other sources.

The A-weighted sound-power data for the tests ranged from 1 to 16 mW. This can be converted to decibels (see appendix A). Peaks in the narrow-band spectra usually occur at frequencies of the orders of excitation of the components of the power train. This can be used to identify noisy components and determine how much the sound power can be reduced by reducing the noise of individual components. Also sources of noise can be identified, using the radiation pattern of sound-power flow on the surface of the measurement hemisphere.

Figure 7.8. Measuring the sound power of light vehicles.

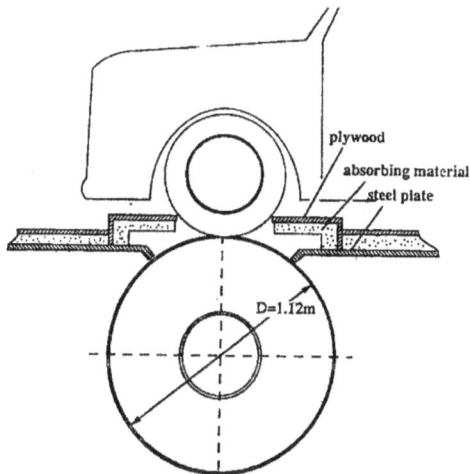

plywood

absorbing material

steel plate

D=1.12m

Figure 7.9. Isolating the noise of dynamometer roll and tires.

With less expensive microphones and a modern signal processing system, it is possible to measure sound power using an array of probes over the measurement hemisphere. Sound power can then be determined in one step. To make it easier to move a source in and out of the measurement hemisphere, sound power could also be determined in two steps with an array over half the hemisphere.

7.4 Sound-power flow of tire interaction with a road surface

The noise of a truck tire can be separated from the other noise sources in the vehicle using the system shown in figure 7.10. This system was used at General Motors many years ago [4], where the fifth wheel of the towing vehicle has to be inverted to keep the single-wheel system upright. Care has to taken when cornering on curved roadways. The noise of the tire emanates principally from the contact patch with the roadway. The type and condition of the road surface plays a major role in generating noise. Sound-power flow is measured using probes shown in figures 1.6 and 1.7, protected from interference by airflow using foam rubber windscreens.

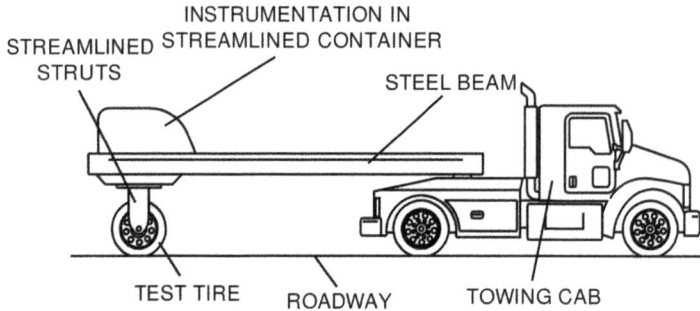

Figure 7.10. Separating the noise of truck tires from other noise sources.

7.5 Railroad noise

Regulations for control of railroad noise are discussed in [5]. These are based on the Code of Federal Regulations section 40, Part 201 and section 49, Part 210. Railroad noise can be of concern in urban communities and to wildlife in forested and open areas.

7.6 Measuring the sound power of some other sources

The semicircular array has also been used for other noise sources [3], for example the garden tractor shown below in figure 7.11. As before, measurement accuracy is tested using a standard sound-power reference source.

Another source is the construction equipment shown below in figure 7.12. This is a standard procedure, which uses two probes on the quadrant of a circle rotated over a hemisphere. However, the procedure of this standard is not an accurate measure of sound power, because there are too few probes on the rotated quadrant.

Measuring the sound power of a vacuum cleaner is illustrated in figure 7.13. Here measurements are made manually at points on a rectangular frame enclosing the cleaner. A windscreen protects the two-microphone probe from noise due to the airflow.

7.7 Gated sound-power measurement

Gating can be used to measure the sound power emitted over different parts of an engine cycle. This was demonstrated for the single-cylinder engine of the garden tractor in figure 7.11. The gating method is described in [3]. Sound power was gated for the intake, compression, combustion and exhaust parts of the engine cycle. As shown in table 7.2, the sum of the contributions of each part of the sequence was found to agree with the overall un-gated sound power, thus confirming the accuracy of the gating procedure. As might be expected, the data shows that the intake and exhaust contribute most noise.

7.8 Measuring the sound power of a moving source

The sound power of a steady-state source moving at constant speed can be measured using a semicircular array of full-space vector probes. The source is assumed to

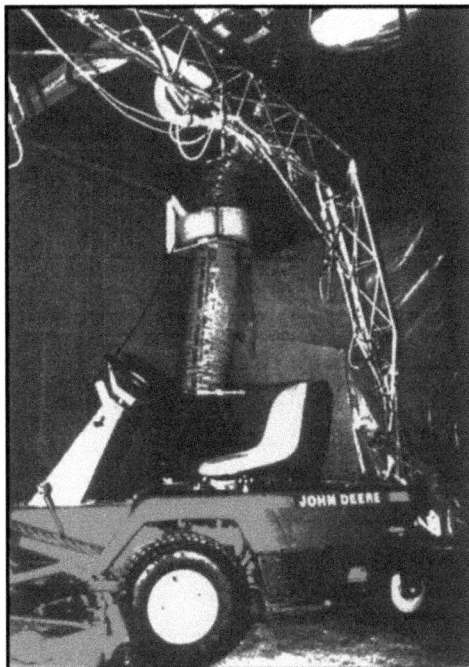

Figure 7.11. Garden tractor.

Figure 7.12. Construction equipment.

move in a straight line along a roadway beneath the arch formed by the array, as shown in figure 7.14. The roadway is assumed to be a rigid surface.

This is equivalent to moving the array over the source at the same speed. As shown by the dashed line in the figure, the enclosing surface is a half cylinder extending in both directions above the roadway. The vector probes track the source. Reflections from objects off the roadway and outside the half cylinder do not contribute to the measurement. Doppler shift and wind noise are included.

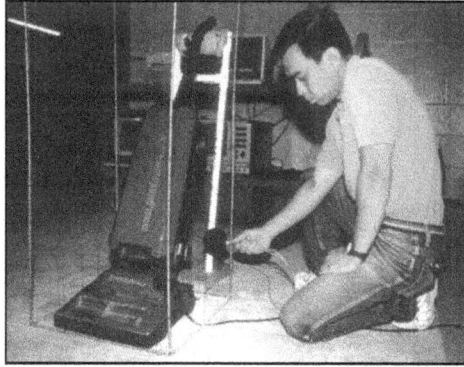

Figure 7.13. Measuring the sound power of a vacuum cleaner.

Table 7.2. Gated sound power for the various parts of the engine cycle.

Gated Sound Power (mW)				Sum of Gated Sound Power	Ungated Sound Power (mW)
Comb.	Exhaust	Intake	Compr.		
1.64	5.03	2.64	1.96	11.27	11.71
1.61	4.91	2.56	1.88	10.96	10.95
1.64	4.67	2.61	1.95	10.87	11.04
1.56	5.02	2.58	1.97	11.13	11.18
Corresponding Sound Power Data in dB(A)					
92.1	97.0	94.2	92.9	100.5	100.7
92.1	96.9	94.1	92.7	100.4	100.4
92.1	96.7	94.2	92.9	100.4	100.4
91.9	97.0	94.1	92.9	100.5	100.5

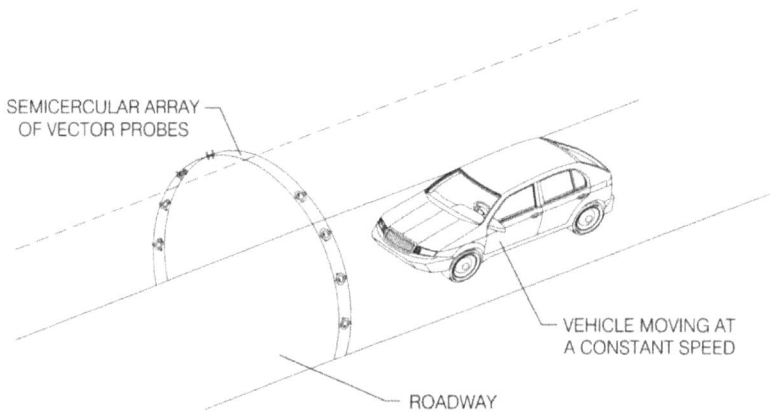

Figure 7.14. Measuring sound power of a moving motor vehicle.

7.9 Sound of agricultural machinery

The sound of agricultural machinery moving over the ground can be measured using a system similar to that in figure 7.14. However, in this case, there is absorption by the ground. The machinery can include tractors, grass mowers, combine harvesters, and sprayers.

7.10 Sound on and in water

Sound on water would include sound from large and small boats and ships in rivers, harbors, and canals. It would be particularly noticeable in the Suez and Panama Canals and in rivers such as the Amazon, Nile, Ganges, and Yang-Tse. Some outboard motors are particularly noisy.

In [6], tests were conducted in water, using a moving source that has a sound field symmetric about its axis. The source was a hydrophone driven in reverse. Because of the symmetry, only one full-space vector probe is required. The source was guided past the probe at a constant speed along a straight line, as shown in figure 7.15, which is the axis of a hypothetical cylinder enclosing the source. The probe tracks the source and measures the component of sound-power flow perpendicular to the cylinder at each position of the source.

Sound power is determined by integrating the measurement over the hypothetical containing cylinder. It is necessary to integrate only along a finite length of the cylinder because measurements perpendicular to the cylinder are negligible when the source is far from the probe. In [4] the sound power of the source was determined both when the source was moving and when it was stationary. The sound power of the moving source was found to be about 14% or 0.5 dB greater than the stationary source. This is probably due to flow noise over the moving source. The method can be used to determine the sound power of a submarine or any other moving underwater source.

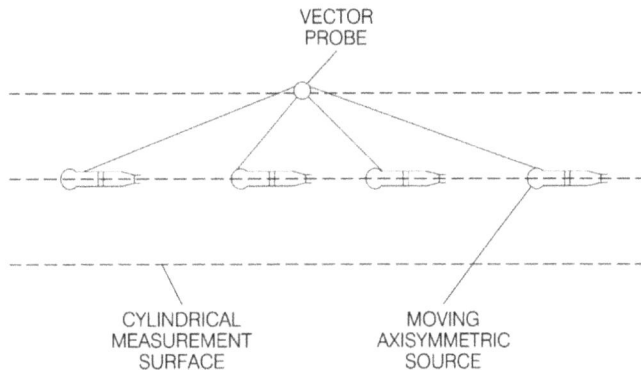

Figure 7.15. Measuring sound power of a moving source in water.

References

[1] Fahy F J 1995 Sound Intensity, 2nd edn, Chapters 4 & 5 (London: Chapman and Hall) p 171

[2] Hickling R, Lee P and Wei W 1997 Investigation of integration accuracy of sound-power measurement using an automated sound-intensity system *Appl. Acoust.* **50** 125–40

[3] Wei W, Hickling R and Lee P 1992 Gated sound-power measurement using an automated sound-intensity system *Noise Control Eng. J.* **39** 17–9

[4] Wilken I D and Hickling R 1976 Measurement of truck tire noise using a single wheel trailer *S.A.E. Symposium on Highway Tire Noise* Paper No. 762014

[5] Hickling R 1997 Surface Transportation Noise ch 88 2 *Encyclopedia of Acoustics* ed M J Crocker (New York: Wiley) pp 1073–81

[6] Wei W and Hickling R 1995 Measuring the sound power of a moving source *J. Acoust. Soc. Amer.* **97** 116–20

Chapter 8

The sound power of a source measured at its surface

8.1 Sampling sound-power flow from a vibrating surface

Sound-power flow can be sampled in small regions of the surface of an engine or other vibrating object using a probe with two pressure sensors, as shown in figure 8.1. The measurement should be made with the probe perpendicular to the surface.

Sound-power flow can be determined for different parts of an object, such as the fuel-injection system in figure 8.2. The relative magnitude of the flow can be shown by color coding.

8.2 Scanning using simultaneous space and time averaging

To obtain a measure of sound-power flow from a large area of a sound source, such as the oil pan of the diesel engine in figure 1.2, requires a different technique. Initially it was thought that this could be determined by dividing the source into components, then making measurements at a number of points on each component and

Figure 8.1. Sampling sound-power flow on a small engine.

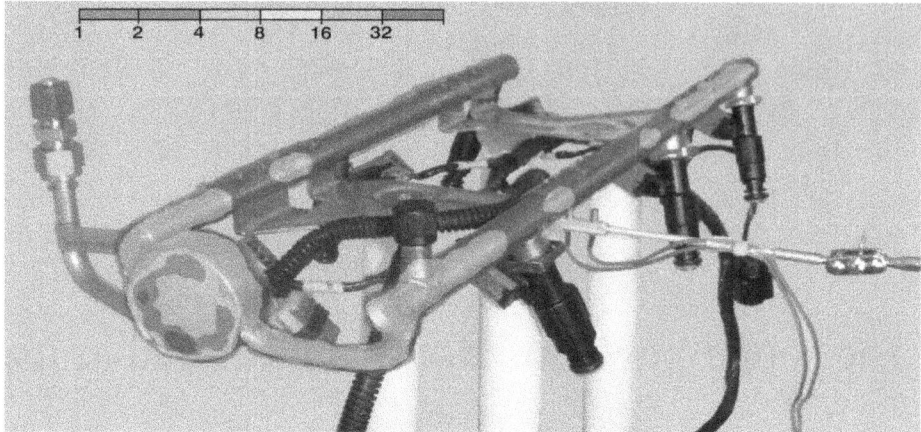

Figure 8.2. Sound-power flow from a fuel injection system.

integrating over the component. The total sound power of the source would then be obtained by summing the sound powers of the components.

Unfortunately this has two drawbacks. These are (a) the complexity of components making it difficult to define their surface geometry and (b) sound-power-flow measurements at the surface of a large component was found to vary erratically when the probe was moved from one point to another. Because of these drawbacks, a different approach was used. This was to scan the surface with a moving probe, simultaneously averaging in space and time [1]. The probe is held perpendicularly to the surface, moving continuously and making sure all parts of the surface are covered equally. This procedure is similar to painting with a brush. It provides a single-number measure of the sound power of the component. Usually the procedure is performed by hand and can be repeated to check for accuracy.

8.3 Sound-power ranking of the components of a source

The sound powers of the components of a diesel engine were determined [1] using the simultaneous space-and-time averaging procedure just described. The results were ranked schematically, as shown in figure 8.3.

The exhaust was directed out of the test area and its noise is not included in the ranking. Also the engine is relatively quiet, because its rotational speed was kept low for persons working in the area. The sum of the sound powers of the components in figure 8.3 is about 0.5 mW. This agrees closely with the sound power that was measured for two kinds of hypothetical surface enclosing the engine: (a) a spherical surface and (b) a set of plane surfaces. The agreement with the first method was within about 2% and within about 12% with the second.

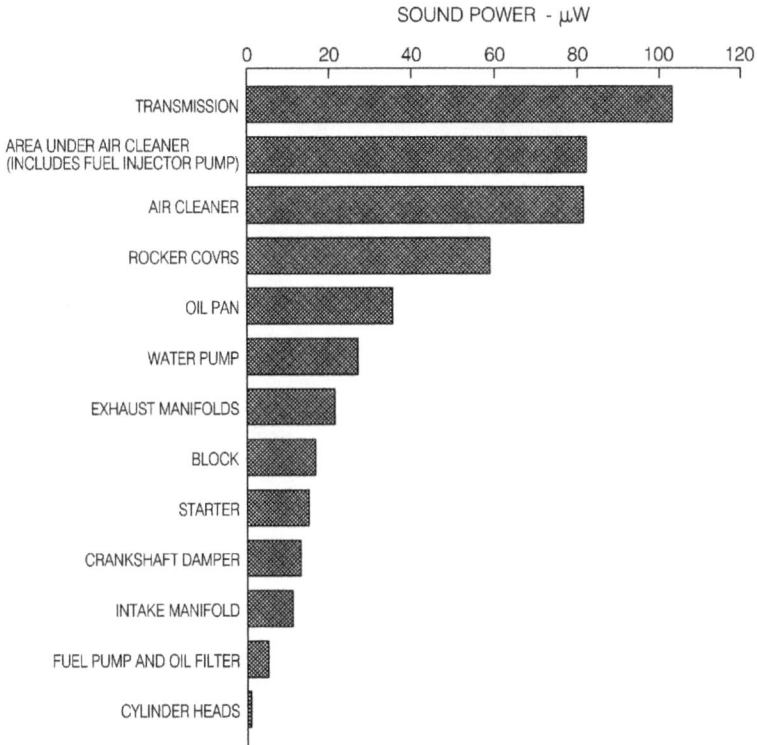

Figure 8.3. Ranking sound power of components of a diesel engine.

Reference

[1] Chung J Y, Pope J and Feldmaier D A 1979 Application of Acoustic Intensity Measurement to Engine Noise Evaluation SAE paper 790502

Chapter 9

Enhancement of sound power by sound from other correlated sources

9.1 Introduction

As is well known, the sound power of a sound source is enhanced by sound from adjacent correlated sources. Correlated means that the sources that are in phase and driven by the same kind of excitation. Examples are: (a) the blending of sound from two loud speakers [1], and (b) the increase in the sound power of a source near a reflecting surface [2]. The correlation enhancement decreases as the sources are moved further apart and as frequency increases.

Before the use of two-microphone probes, the sound powers of the different components of a complex source, such as an engine, were measured using lead wrapping. Several diesel engines were tested in this way [3]. In these tests, the engine is wrapped with lead sheet, with a thin layer of soft absorbing material between the lead sheet and the engine. The lead wrapping is removed exposing each component one at a time. The sound power of each unwrapped component is then measured over a hypothetical surface enclosing the engine. The wrapping is then replaced and the wrapping of another component removed. The sound powers of the components are then ranked and summed. The sum is compared to the sound power of the completely unwrapped engine. As shown in table 9.1, the sound power of the unwrapped engine for three different diesel engines is invariably greater than the sum of the individually unwrapped components. This is because lead wrapping prevents acoustic interaction between the correlated components of the engine.

The interaction between correlated components was then modeled [4] using the surface vibrations of a sphere. The sound power of a component of the surface was calculated by integrating sound-power flow over the surface of the component in two ways. First, to simulate lead wrapping, the other components of the surface were assumed to have zero sound-power flow. Second, the other components were assumed to have sound-power flow correlated with the particular component being

doi:10.1088/978-1-6817-4453-7ch9

Table 9.1. Sound power for three different diesel engines, unwrapped and lead wrapped.

Engine	Operating condition	(U) Sound power unwrapped engine dB (A)	(S) Sum of components using lead wrapping dB (A)	(U–S) Difference dB(A)	Corresponding sound power ratio
A	rated speed, 1/2 load	107.8	106.1	+1.7	1.48
B	rated speed, 1/2 load	110.2	108.1	+2.1	1.62
C	rated speed, 1/2 load	107.0	105.2	+1.8	1.51

calculated. This created an enhancement of the sound power of the component. For spheres the size of truck diesel engines the effect of the other correlated components is in the range shown in table 9.1. The model shows that enhancement increases with decrease in both the source size and the maximum frequency of interest. The enhancement effect is thus greater for smaller engines.

The sound power of the drive train of a vehicle is increased by reflection from the underlying roadway while, on the other hand, there is no enhancement between correlated sources that are a distance apart, as in the case of the two correlated standard reference sources in the reverberation room in figure 7.6, where one source is inside the enclosing measurement hemisphere and the other is outside, in the corner of the room.

References

[1] Wolff I and Malter L 1929 Sound Radiation from a System of Circular Diaphragms *Phys. Rev.* **33** 1061–5
[2] Ingard K U and Lamb G L 1957 Effect of a Reflecting Plane on the Power Output of Sound Sources *J. Acoust. Soc. Amer.* **29** 743–4
[3] Trella T, Mason R and Karsick R 1979 External Surface Noise Radiation Characteristics of Truck Diesel Engines, their Far-Field Signatures and Factors Controlling Abatement SAE paper Number 780174 *Trans. Soc. Auto Eng* 673–701
[4] Hickling R and Marin S P 1988 Enhancement of the Sound Power of a Component of a Complex Noise Source by Sound from Other nearby Components *J. Acoust. Soc. Amer* **84** 262–74

Chapter 10

Transfer-function methods applied to incident and reflected components of sound-power flow in a duct

10.1. Measuring normal-incidence absorption with an impedance tube

Sound-power flow in an impedance tube, with an acoustic driver at one end and a test sample of absorbing material at the other end is illustrated diagrammatically in simple form in figure 10.1. The flow has incident and reflected components, indicated by the arrows. These components can be measured and separated from each other using a transfer-function method, with two microphones flush with the tube wall. Transfer functions were defined previously in chapter 2. The transfer-function equations used here are given in standards and in journal publications [1–4]. The

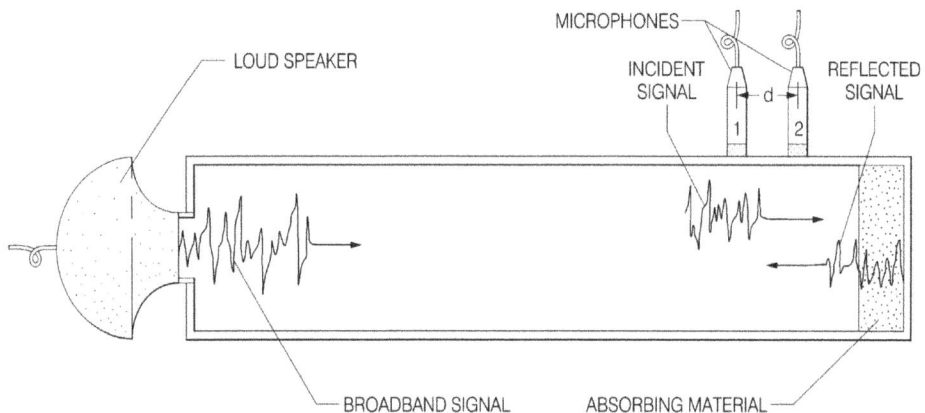

Figure 10.1. Sound-power flow in an impedance tube.

required procedure is explained in detail in [1]. An additional microphone is often used to extend the frequency range of the measurements.

Some details are shown in the photograph in figure 10.2.

Instrumentation for measuring normal-incidence absorption is available from different manufacturers, replacing the previous standing-wave ratio method. This instrumentation can also measure transmission loss across an acoustical element, such as a silencer, using two pairs of microphones, one on each side of the element. Special switching procedures [1, 2] can be used to match the response of the microphones in each pair over the frequency range of the measurements. In addition, there is a standard method for measuring absorption due to random-incidence sound in a reverberation room [4].

10.2. Fluid flow superimposed on sound-power flow in a duct

Fluid flow is added [5] to the measurement of transmission loss in ducts to provide for laboratory testing and design of engine silencers. Figure 10.3 presents a diagrammatic view of the incident and reflected components of sound-power flow with fluid flow superimposed.

Sound-power flow for plane waves in a duct *without* fluid flow is calculated using the expression

$$I_{av} = (1/\rho_0 c)\text{Im}[S_2(f) \cdot S_1^*(f)]/\sin(kd) \qquad (10.1)$$

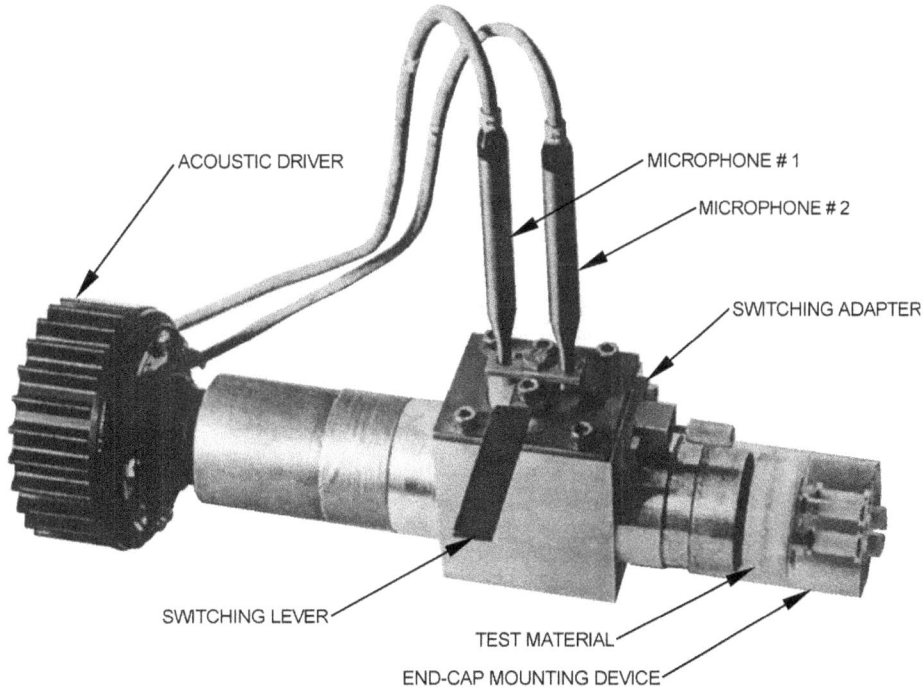

Figure 10.2. Photograph of the impedance-tube system.

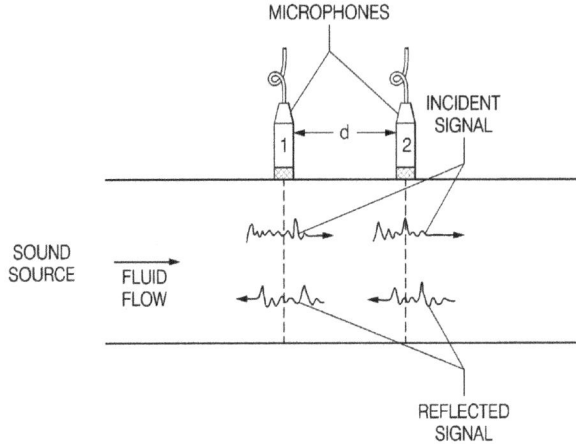

Figure 10.3. Sound-power flow with fluid flow.

This is different from the three-dimensional cross-spectral formulation for sound-power flow given in equation (1.15), which is derived using finite-difference approximations. These approximations require the microphone spacing d to be small, as indicated previously in equation (1.21). In contrast the spacing d between the microphones in equation (10.1) does not have this restriction.

References

[1] ASTM Standard E1050-08 2008 *Impedance and Absorption of Acoustical Materials Using a Tube, Two Microphones and a Digital Frequency Analysis System* American Society for Testing and Materials

[2] Chung J Y and Blaser D A 1980 Transfer-Function Method of Measuring In-Duct Acoustic Properties *J. Acoust. Soc. Amer.* **68** 907–21

[3] ISO Standard 10534-2 Acoustics 1998 Determination of Sound-Absorption Coefficient and Impedance in Impedance Tubes — Part 2: Transfer Function Method

[4] ISO Standard 354:2003-Acoustics 2003 Measurement of sound absorption in a reverberation room

[5] Chung J Y and Blaser D A 1980 Transfer-function method of measuring acoustic intensity in a duct system with flow *J. Acoust. Soc. Amer.* **68** 1570–7 **Note error** in Eq. (35): (1-M) in denominator should be (1+M)

Chapter 11

Determining the direction of a sound source in fluids

11.1 Determining the direction of a sound source in air using a vector probe

The direction of a sound source in air is readily determined using the full-space vector probe in figure 1.6. In the tests [1], the source is moved to a number of different positions with known angular directions of azimuth and elevation relative to the probe. The direction to the source is then measured using the probe. Figure 11.1 shows the small scatter in the measured data relative to the known different positions. Agreement with the known positions is within a few degrees.

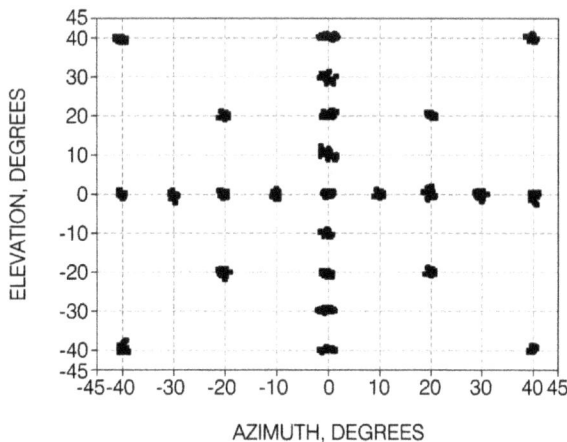

Figure 11.1. Direction of same sound source at different locations.

doi:10.1088/978-1-6817-4453-7ch11

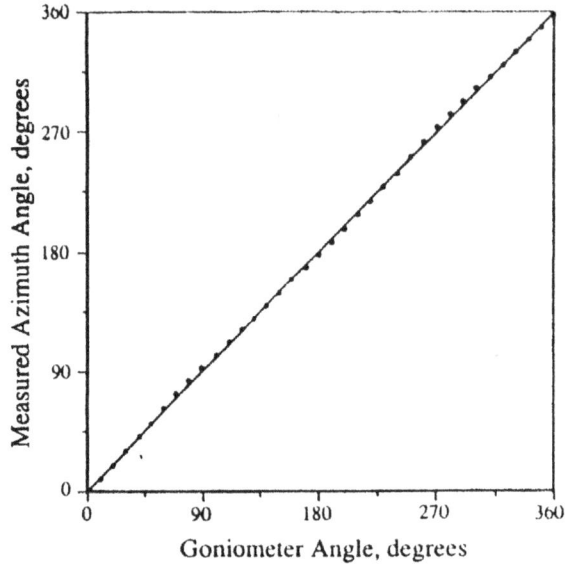

Figure 11.2. Direction of sound source in water.

11.2 Determining the direction of a sound source in water

Determining the direction of a sound source in water is described in [2], using the arrangement of hydrophones in the probe shown in figure 1.8. In the tests, the source is a driven hydrophone emitting 0.5 ms tone bursts. The signals were gated to receive only direct sound and not the reflections from the walls of the tank. The probe was attached to a rod protruding from the center of a goniometer, with the probe in the same plane as the source. The probe was rotated by the goniometer so that it measured the azimuthal angle to the source. The azimuth angle measured by the probe and the goniometer angle are compared below in figure 11.2. It is seen that the data is accurate to within ±2°. The undulations in the data suggest this small error may be due a structural inaccuracy in the fixtures.

11.3 Echolocation in sediment-laden water

Water is frequently laden with sediment, making it is difficult to use vision or illumination to detect objects, even at short distances. Instead a pulsed source can be used, combined with a vector probe for measuring sound-power flow, as shown schematically below in figure 11.3. The location of an object is determined by its distance and direction. Distance is calculated from the time of flight of the echoes of the sound pulses. Direction is determined from the sound-power flow measured by the vector probe. To penetrate sediment-laden water, the frequency of the pulses has to satisfy equation (1.10).

This device can be attached to a diver's arm. Several objects can be detected at one time within the beam of the source, as shown in figure 11.4.

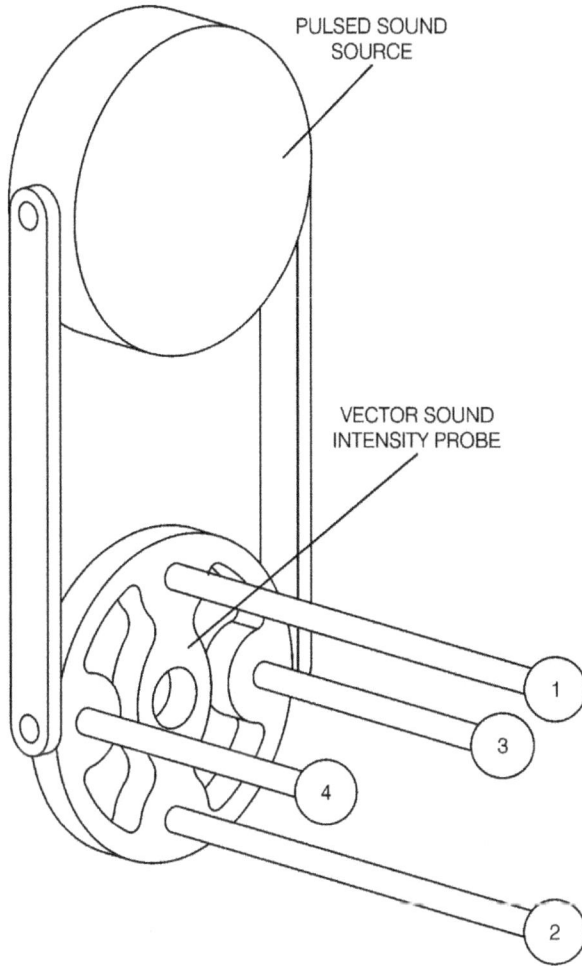

PULSED SOUND
SOURCE

VECTOR SOUND
INTENSITY PROBE

1

3

4

2

Figure 11.3. Combination of sound source and vector probe.

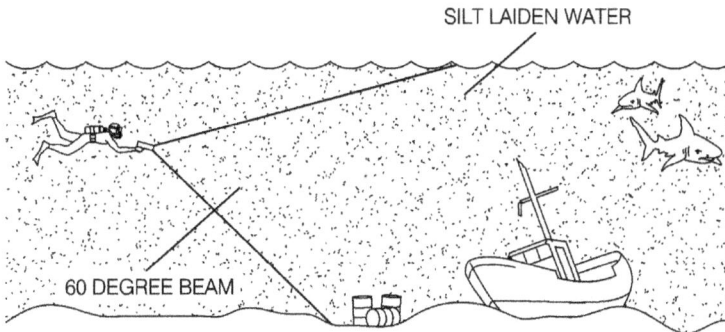

SILT LAIDEN WATER

60 DEGREE BEAM

Figure 11.4. Diver using a combination of source and vector probe.

11.4 Forward-looking sonar for ships and boats

Another application that uses a combination of a sound source and a half-space vector probe is the forward-looking sonar system for ships and boats [3]. This is illustrated in figure 11.5. This system could have prevented many accidents, including the well-known wreck of the liner Costa Concordia.

11.5 Directional hearing of humans and mammals

Rayleigh [4, 5] was the first to provide a theory explaining directional hearing. He assumed the head is a rigid, solid sphere and he postulated that the direction of sound is detected, at low frequencies, using phase differences between the ear positions, and, at high frequencies, by amplitude differences, as illustrated below in figure 11.6.

The separation into low and high frequencies is called the duplex theory. More recent versions [6] of duplex theory use the terms, ITD and ILD, where ITD is the low frequency, interaural time difference and ILD is the high frequency interaural pressure-level difference. Standards [7] have been established based on duplex theory.

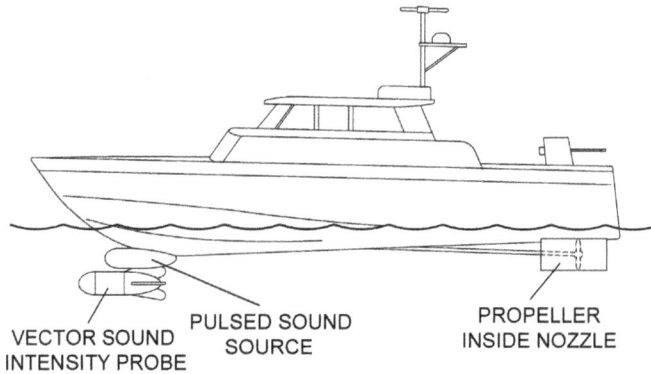

VECTOR SOUND INTENSITY PROBE   PULSED SOUND SOURCE   PROPELLER INSIDE NOZZLE

Figure 11.5. Schematic of a forward-looking sonar system.

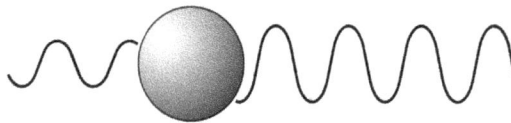

(a) HIGH-FREQUENCY SHADOWING

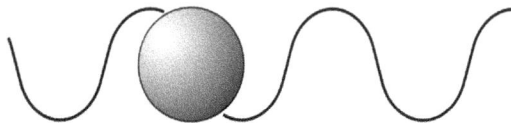

(b) LOW-FREQUENCY PHASE DIFFERENCE

Figure 11.6. Rayleigh's duplex theory.

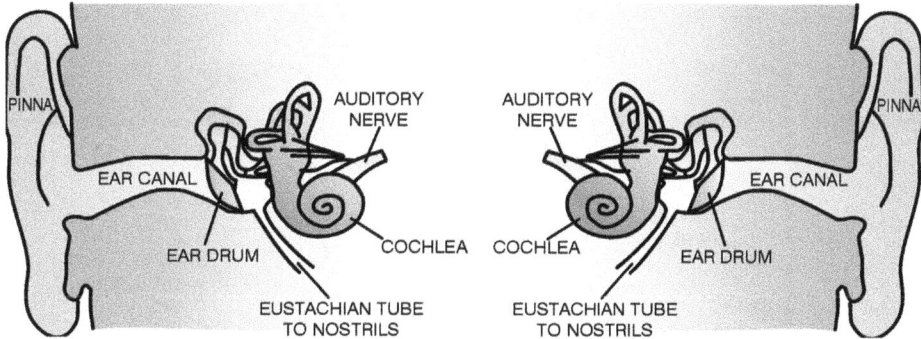

Figure 11.7. Human hearing system.

The main deficiency of duplex theory is the assumption of a solid head. Obviously the head is not solid. The human hearing system inside the head is as shown in figure 11.7. Another problem is the division of the frequency range of directional hearing into two parts, one at high frequencies and the other at low frequencies. Moore [8] discusses the error that can occur with this separation. A more realistic explanation is required, based on an actual head and brain. Such an explanation would apply, in general, to mammals, in air and in water, as well as to humans.

How the brain accomplishes directional hearing, using inputs from the auditory nerves is unknown. The auditory area of the brain performs a variety of functions and is quite complex [9, 10]. Input from both ears is transmitted into each half of the brain, as shown below in figure 11.8 [10].

Even greater complexity of the auditory area of the brain has been indicated in other studies [11].

11.6 Possible use of vector sound-power flow for finding direction by the brain

As shown previously in figure 1.5, vector sound-power flow is determined most simply using four acoustic sensors in the regular tetrahedral arrangement. Thus it does not seem unreasonable to assume that the brain of humans and mammals can use this basic, simple method for directional hearing, since there appears to be a sufficient number of possible acoustic terminal nodes in the brain to create a regular tetrahedron for determining the direction of vector sound-intensity by the brain. The sound pressures at the midpoints are determined using finite-difference approximations. The approximations are valid if the distance d, along a tetrahedral edge between any two sound-pressure sensors, satisfies the relation $d \ll \lambda/2\pi$, where λ is the wavelength of sound. The wavelength λ is equal to the speed of sound c in the medium, divided by f the acoustic frequency. Hence the upper limit for d is given by

$$d \ll c/(2\pi f) \tag{11.1}$$

The frequency range of human hearing extends approximately up to 20 kHz. Also the brain can be assumed to consist of water-like tissue with $c = 1540$ m s^{-1} [12]. It

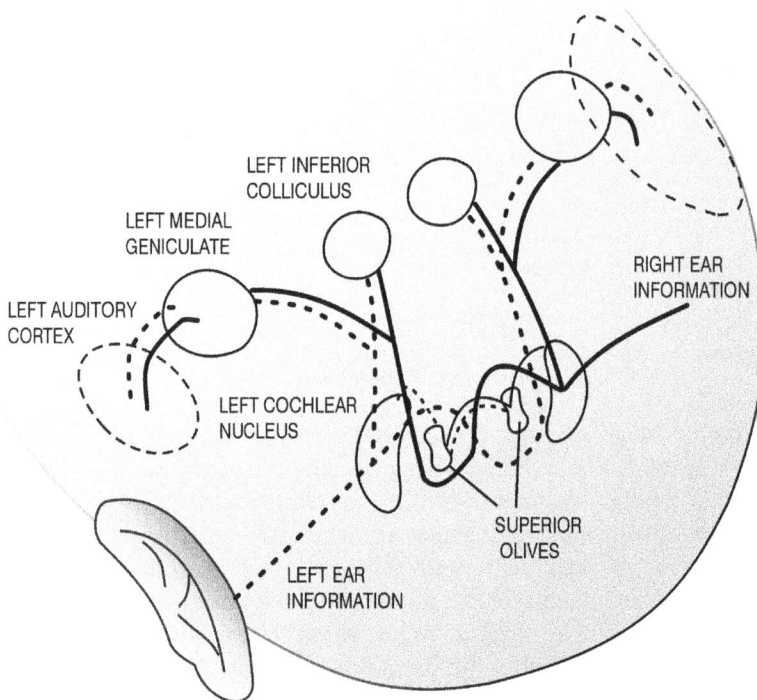

Figure 11.8. Auditory area of the brain.

follows that *d* is about 12.8 mm. For adult humans, the brain has a cross-section of roughly 100 mm. Hence the dimension *d* can readily fit in the brain. Children have smaller heads and are more receptive to high-frequency sound from women and other children. For them, *d* is slightly smaller. For mammal heads in general, the distance *d* varies with the size of the head. Bats have very small heads [13–15], allowing them to operate at ultrasonic frequencies, while the large brain of elephants allows them to respond to low-frequency infrasound over long distances [16]. Dolphins and other sea mammals are known to have large brains for directional hearing in water.

11.7 Tests of directional hearing using a two-dimensional array of speakers

The tests were conducted in an anechoic chamber at the General Motors Proving Ground at Milford, Michigan, using a number of young subjects whose age was less than 30. These were either permanent or temporary employees. Because of their youth and from their own statements, it was decided that none had hearing disabilities. There were about eight male and the same number of female subjects, the females being generally slightly younger than the males. The test area in the anechoic chamber is shown in figure 11.9. The arrangement of speakers is shown in figure 11.10.

Figure 11.9. Directional hearing tests in an anechoic chamber.

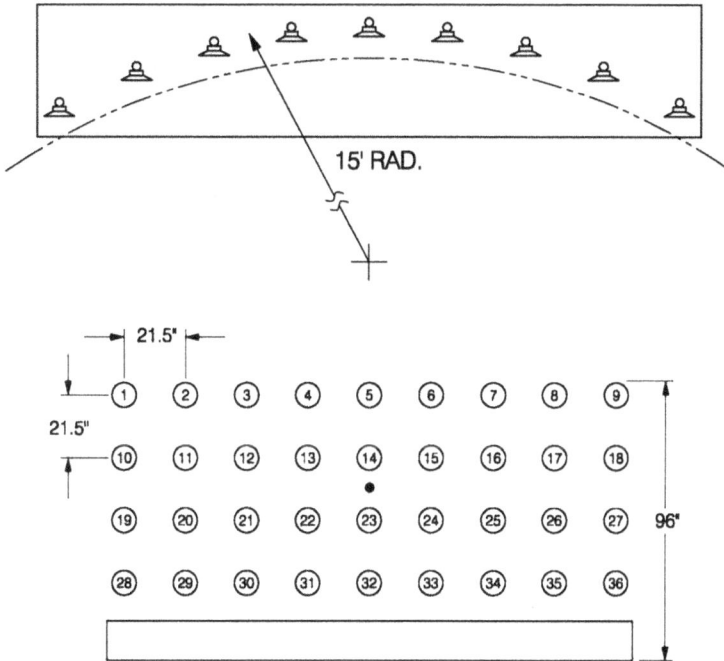

Figure 11.10. (a) Plan view of speakers. (b) Vertical view of speakers.

The subjects were tested in the dark, with their heads pointing forward, looking at a visible laser spot in the center of the array, but without prior knowledge of the location of the driven speaker. The subjects then noted their impression of the location of the speaker in the array, using a light temporarily lit on their desk. The results shown here are for speakers driven only by short bursts of sound, such as the sudden fracture of a machine part or gunfire. It was found that there was some scatter in the accuracy of the data, with the concentration being mainly one speaker away from the actual source. There was no difference between male and female subjects.

11.8 Insect sounds

11.8.1 Near-field acoustic directional sensing by ants

Sound-power flow is very different in the near-field region of an ant [17]. It is known that ants generate sound by stridulation using parts of the extra-skeletal abdomen shown in figure 11.11. On the top of the gaster, next to the postpetiole, is a set of ridges. A scraper on the postpetiole moves back-and-forth over the ridges generating sound as the gaster moves up and down.

The process is similar to generating sound with a washboard. Typically the sound has a frequency of about 700 Hz [17], which can be heard faintly by holding a fire ant close to the ear in a quiet environment. Recordings of the sounds are provided on the web [18]. The wavelength of the stridulation sound is about 0.2 m. This is much larger than the size of an ant, which typically is about 2–3 mm long. A detailed analysis of the sound field of an ant is given in [17]. The following is a brief summary of this analysis.

The ant's stridulation sound radiates principally from the gaster. The particle velocity of the sound predominates in the vicinity of the ant dropping off rapidly as $1/r^2$ up to a distance r of about 100 mm. This region is called the near field. Beyond this point is the far field where the sound particle velocity drops off as $1/r$. On the other hand, sound pressure drops off as $1/r$ in both the near and far fields.

The $1/r$ dependence in the near field makes differences in sound pressure difficult to detect for an object the size of an ant. On the other hand the much steeper $1/r^2$ dependence of sound-particle velocity allows an ant to detect acoustic signals from another ant using the concentration of hair sensilla at the ends of its antenna. The hair sensilla are shown in figure 11.12. The use of hair sensors to detect sound is known by entomologists, as cited in [17]. The joints of the antennae are flexible allowing an ant to position the hair sensors so that acoustic signals in the near field are maximized. Adjusting the position of the antennae also allows an ant to determine the direction of these signals. Hence an ant can communicate with other ants and locate other small sources of sound within the near field. On the other hand ants appear to be deaf to sound in the far field. This is easily verified experimentally, because sounds originating from a person have little or no effect on ants.

Ants are a favorite topic in movies, novels, and even in music. However, apart from using their stridulation sounds [18] in music, these representations have not been very realistic. Understanding of ant sounds has been neglected by

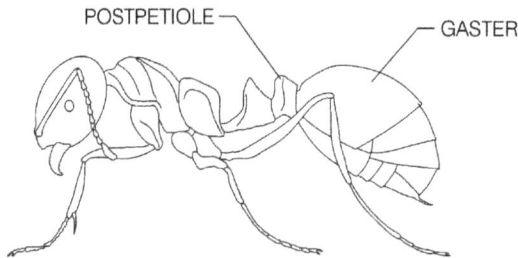

Figure 11.11. Extra-skeletal structure of a fire ant.

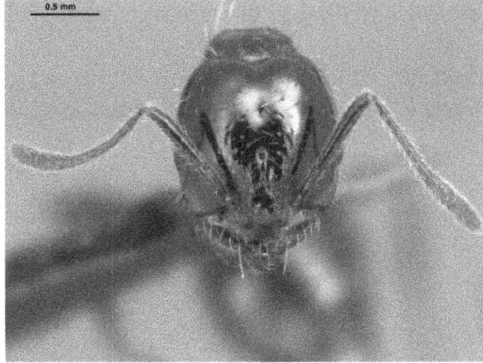

Figure 11.12. Antennae of a fire ant.

myrmecologists [19], who have tended to ignore sound and to emphasize chemical communication by ants. Formic acid is believed to be a significant contributor.

References

[1] Hickling R and Brown A W 2011 Determining the direction of a sound source in air using vector sound-intensity probes *J. Acoust. Soc. Amer.* **129** 219–24

[2] Hickling R and Wei W 1995 Use of pitch-azimuth plots in determining the direction of a sound source in water with a vector sound-intensity probe *J. Acoust. Soc. Am.* **97** 856–66

[3] Hickling R 2012 Forward-Looking Sonar for Ships and Boats *US Patent No.* 8,203,909

[4] Rayleigh L 1945 *The Theory of Sound* vol 2 (New York: Dover) pp 442–3

[5] Rayleigh L 1913 On our perception of sound direction *Phil. Mag.* **6** 214–32

[6] Kuhn G F Physical acoustics and measurements pertaining to directional hearing ch 1 ed W A Yost and G Gourevitch *Directional Hearing* (London, New York: Springer)

[7] Standards, ANSI S3.36/ASA58-1985; IEC 60959

[8] Moore B C J 2012 *An introduction to the Psychology of Hearing* 6th Edn (London: Emerald) p 255

[9] Casseday J H and Covey E C Central Auditory Pathways in Directional Hearing pp 107–45 Directional Hearing Part 2, ch 5 ed W A Yost and G Gourevitch (New York: Springer)

[10] 1987 http://hyperphysics.phy-astr.gsu.edu/hbase/sound/anerv.html Date last viewed 10/15/14

[11] Papo D, Buldu J M, Boccaletti S and Bullmore E T 2014 Complex network theory and the brain *Philosophical Transactions of the Royal Society of London, B: Biological Sciences* **369**

[12] Ludwig G D 1950 The Velocity of sound through tissues and the acoustic impedance of tissues *J. Acoust. Soc. Am.* **22** 862–6

[13] Simmons J A 2002 Directionality of Biosonar Broadcasts and Reception by the Ears Tutorial Lecture, ASA Spring Meeting

[14] Covey E 2005 *Neurobiological Specializations in Echolocating Bats* (New York: Interscience) pp 1103–36

[15] Griffin D R 1958 *Listening in the Dark-The Acoustic Orientation of Bats and Men* (New Haven: Yale University Press) total pages 413

[16] Payne K 1999 *Silent Thunder: In the Presence of Elephants* (London and New York: Penguin Books) total pages 284

[17] Hickling R and Brown R L 2000 Analysis of acoustic communication by Ants *J Acoust. Soc. Am.* **108** 1920–9

[18] www.olemiss.edu/~hickling/

[19] Holldobler B and Wilson E O 1990 *The Ants* (Cambridge, MA: Harvard University Press) p 732

Sound-Power Flow
A practitioner's handbook for sound intensity
Robert Hickling

Chapter 12

Computer-visualation of the free vibrations of metal spheres and shells

12.1 Introduction

Calculations have been performed to visualize the modes of free vibration in the interior of metal spheres and spherical shells [1]. These are based on the method of separation of variables, which can be applied to a number of simple body shapes [2]. More general shapes can be treated using finite-element techniques, such as NASTRAN [3]. The vibrations of complex structures such as an engine or aircraft, will have similar characteristics as those in the simple body shapes, but will be distributed in a more complex manner. Experimental modal analysis [4] shows the individual modes of vibration on the surface of a structure. However it does not provide information about sound-power flow in the interior.

The free modes of vibration of a solid sphere depend on the size and material of the sphere as well as the frequency of the modes. Table 12.1 provides a list of the axisymmetric modes of a solid aluminum sphere, as a function of the non-dimensional frequency ka, where a is the radius of the sphere. The modes are used later to describe the response of the sphere to incident sound in water. Here k is derived from $k = 2\pi f/c$ where f is frequency and c is the speed of sound in fresh water, 1440 m s^{-1}. The elastic constants of aluminum are assumed to be as follows: density $\rho = 2.965 \times 10$ kg m^{-1}; Lamé constant $\lambda = 61$ GPa and shear constant $\mu = 25$ GPa.

The first row in table 12.1, for the parameter $n = 0$, is for the spherical or breathing modes of vibration and are not discussed here. The missing fundamental at $(0, 1)$ is a rigid-body oscillation back and forth along the axis of the sphere. This can only be excited by an external force and is not a free vibration. The dominant modes are the fundamentals in the first column for n equal to and greater than 2. The remaining corresponding columns to the right are for higher harmonics.

Figures 12.1 and 12.2 show the sound-power flow for the fundamentals at $n = 3$ and 4, for $ka = 8.371$ and $ka = 10.749$ respectively.

doi:10.1088/978-1-6817-4453-7ch12   12-1

Table 12.1. Axisymmetric modes of free vibration of an aluminum sphere.

n	l	0	1	2	3	4	5
0		12.442	27.337	41.570	55.683		
1			7.691	15.436	19.024	22.725	29.405
2		5.610	10.896	18.350	24.218	26.278	
3		8.371	14.307	21.228	27.903		
4		10.749	17.692	24.107			
5		12.974	20.961	27.011			
6		15.124	24.083	29.948			
7		17.230	27.057				
8		19.310	29.898				
9		21.371					
10		23.420					
11		25.458					

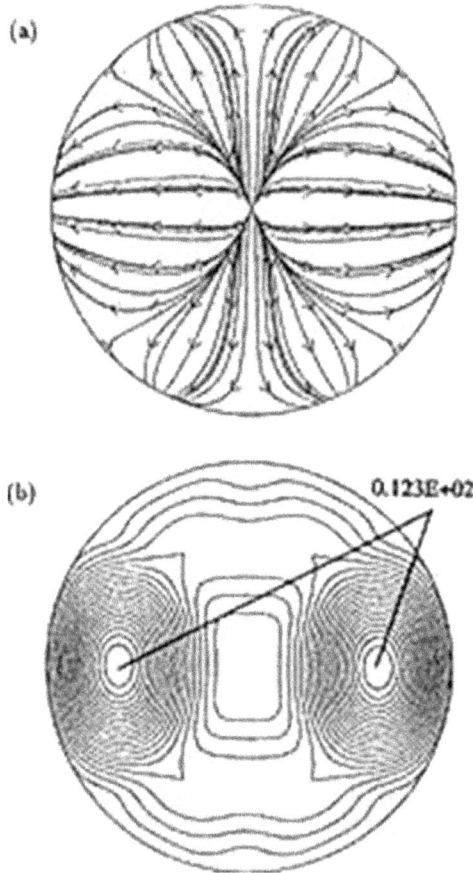

Figure 12.1. Sound-power flow for the fundamentals at $n = 3$.

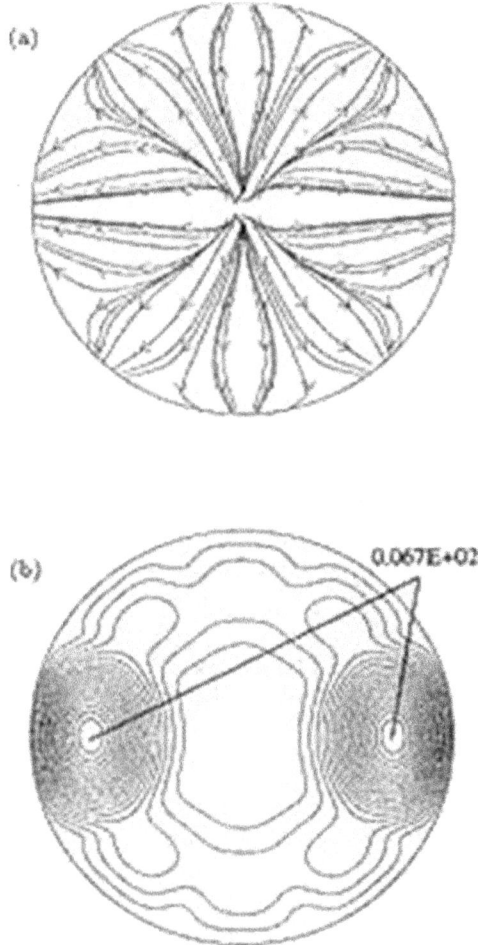

Figure 12.2. Sound-power flow for the fundamentals at $n = 4$.

The directions of the sound-power flow lines are shown in figures 12.1(a) and 12.2(a). These are between the center and the ends of the axis of symmetry at $0°$ and $180°$ and between the center and the annular positions on the surface of the sphere. The number of annular positions increases as the parameter n increases. The direction of the sound-power flow lines is either completely outwards from the center or completely inwards, changing direction four times during a cycle of the resonant frequency. Figures 12.1(b) and 12.2(b) show the contours of the amplitude of sound-power flow. The contours are shown in 20 equal steps between the maximum value and zero. The pattern of the contours is fixed. The amplitude is zero throughout the sphere, when the flow changes direction.

Figure 12.3 shows the sound-power flow for the first harmonic at $n = 3$, $l = 1$, $ka = 14.307$, and figure 12.4 shows the corresponding displacement flow lines calculated using equation (3.2). As before, figures 12.3 (a) and 12.4(a) indicate the direction of flow and figure 12.3 (b) and 12.4. (b) indicate amplitude contours. Note the circular

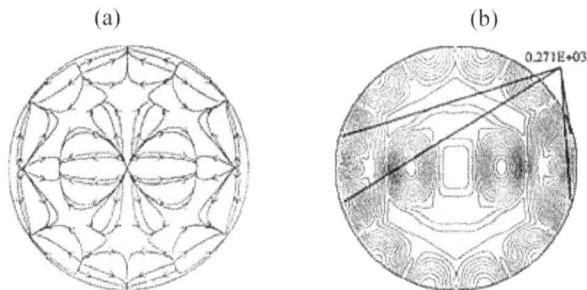

Figure 12.3. Sound-power flow for the first harmonic at $n = 3$.

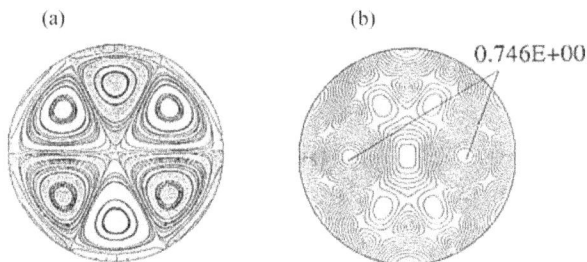

Figure 12.4. Corresponding displacement flow lines.

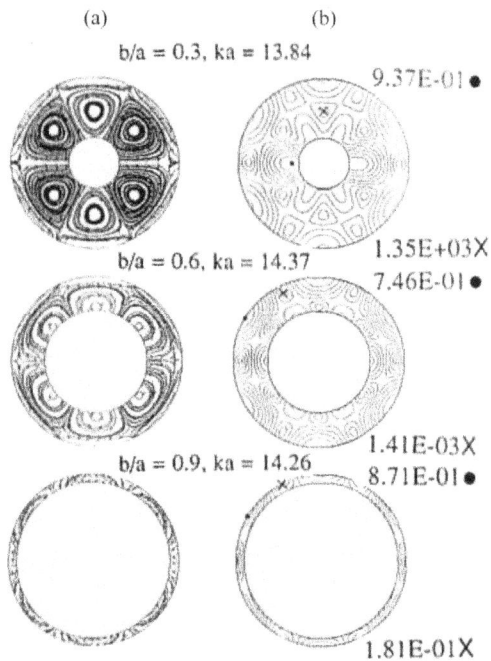

Figure 12.5. Sound-power flow in aluminum shells of decreasing thickness.

flow lines of the displacement vector in figure 12.4. This is a major feature of resonant behavior.

Figure 12.5 shows the displacement vector for the (3, 1) mode and the change in resonant frequency, as the solid sphere in figure 12.5 changes into shells of decreasing thickness, indicated by b/a where b is the inner radius and a is the outer radius of the shell.

As before, the flow lines of the displacement vector in figure 12.5(a) follow curved paths, with the inner surface advancing into the flow-line pattern of the solid sphere as the shell thickens. The positions of the maximum and minimum values of the displacement amplitude in figure 12.5(b) are indicated in the figure by the spot • and letter X respectively.

References

[1] Hickling R 1994 Visualization of elastic vibrations in solid structures, *Comput. Syst. Eng.* **5** 27–40

[2] Morse P M and Feshbach H 1953 Methods of Theoretical Physics (New York: McGraw-Hill) pp 141–3

[3] NASTRAN Finite-Element Analysis Computer-Aided Software. One of the principal suppliers is the MSC Corporation

[4] Allemang R J 1994 Analytical and Experimental Modal Analysis University of Cincinnati Structural Dynamics Research Laboratories, CN-20-263-662

Chapter 13

Computer visualation of acoustic scattering and response of spheres in water

13.1 Introduction

Figure 13.1 shows a solid aluminum sphere being excited by incident plane waves whose frequency is the same as the fundamental mode $ka = 10.749$ in figure 12.2. The interior of the sphere shows only the sound-power flow lines. The outside shows the incident and scattered sound as it progresses from left to right in steps of 45° for half the frequency cycle. The shaded regions indicate positive sound pressure and non-shaded regions indicate negative sound pressure. In the second part of the cycle the direction of the sound-power flow lines inside the sphere is reversed but the same pattern is repeated at each 45° step.

The response of spherical aluminum shells to incident plane waves in water is described in [1], where the shells are assumed to be empty.

Reference

[1] Hickling R, Burrows R K, Ball J F and Petrovic M 1991 Power flow for sound incident on a solid aluminum sphere in water *J. Acoust. Soc. Amer.* **89** 2509–2518

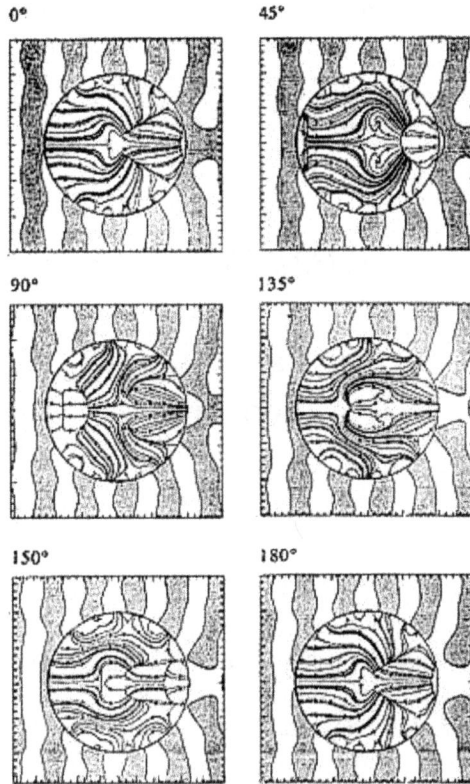

Figure 13.1. Solid aluminum sphere excited by plane waves.

Sound-Power Flow
A practitioner's handbook for sound intensity
Robert Hickling

Chapter 14

Acoustic scattering and focusing by spheres in water

14.1. Introduction

Figure 14.1 shows the sound-power flow around and through an aluminum sphere in water as a function of frequency ka of incident plane waves [1]. At $ka = 0.1$ the sphere has rigid-body motion back and forth in motion with the wave. At $ka = 2.0$ the incident sound-power flow does not yet penetrate the sphere but flows around it creating a narrow shadow zone behind. At $ka = 3.1$ the sound-power flow has started to penetrate the sphere, mainly flowing in and out at the front of the sphere.

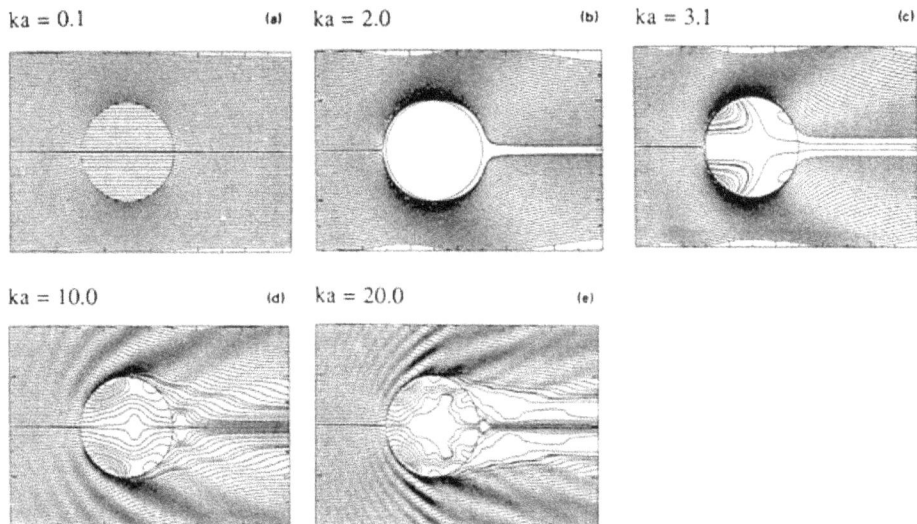

Figure 14.1. Scattering and focusing by an aluminum sphere in water.

doi:10.1088/978-1-6817-4453-7ch14   14-1   © Morgan & Claypool Publishers 2016

ka=5   ka =10

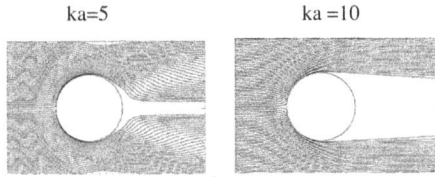

Figure 14.2. Formation of the shadow zone with increasing frequency.

ka = 80

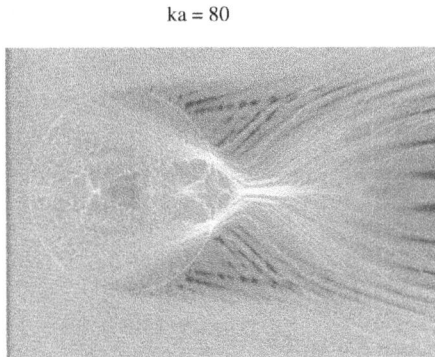

Figure 14.3. Focusing for a lucite sphere in water.

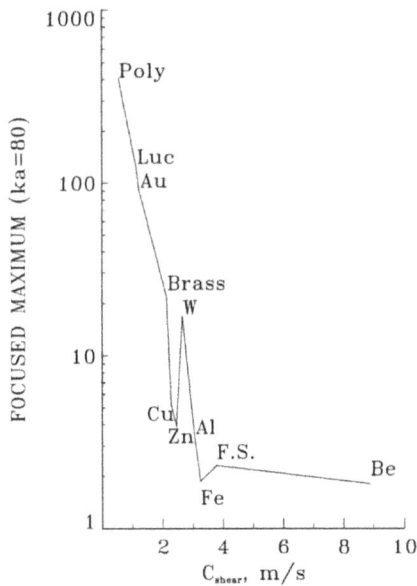

Figure 14.4. Focusing for different materials.

At $ka = 10.0$ and 20.0 sound-power flow passes through the sphere and begins to form a focal region.

If the sphere is rigid and immovable [2], sound does not penetrate the sphere and there is no focusing. A shadow zone forms behind the sphere, which grows as frequency increases, as shown below in figure 14.2.

Focusing is further demonstrated in figure 14.3 for a lucite sphere in water.

The focusing is seen to be quite strong. A lens shape made of lucite, with the same frontal radius a, would produce focusing similar to that in figure 14.3, for a frequency of $ka = 80$.

The maximum spherical focusing for $ka = 80$ for a number of different materials is shown in figure 14.4.

It is seen that lucite is the most practical material for use in water.

References

[1] Hickling R, Burrows R K, Ball J F and Petrovic M 1991 Power flow for sound incident on a solid aluminum sphere in water *J. Acoust. Soc. Amer.* **89** 2509–18
[2] Hickling R and Wang N M 1966 Scattering of sound by a rigid movable sphere *J. Acoust. Soc. Am.* **39** 276–9

Chapter 15

Computed results compared with experimental data

15.1. Introduction

There are a number of ways of making numerical predictions, some of which have been mentioned here. They originate as mathematical equations, which can sometimes be solved in closed form when the equations and boundary conditions are relatively simple. For more complex shapes, it is necessary to use computer codes such as NASTRAN [1] and computational fluid dynamics. The solutions then have to be presented so that they are easily understood visually. Of course, the original equations, the principles on which they are based and the computational methods have to be accurate. Finally and perhaps most importantly the results have to agree with observation.

15.2. Comparison of theory and experiment for scattering by aluminum spherical shells

In sonar detection it was originally believed that solid objects responded to incident sound as though they were rigid bodies [2]. In particular the multiple-echo effect observed with spherical targets was originally believed to be due to reflection from different parts of the sphere's surface. It was only later [3] that it was shown to be due to the elastic response of the sphere.

Tests were conducted [4] to compare theory with experiment for aluminum spherical shells in water. The tests were conducted under a barge in a fresh-water lake, as shown in figure 15.1.

In the tests, the backscattered-echo amplitudes are shown in figure 15.2. These were both calculated and measured for an aluminum shell, 5 inches in diameter with $b/a = 0.95$ and filled with water.

The results are compared in figure 15.2 as a function of frequency ka where the solid line represents computed and the dashed line measured data. It is seen that the agreement is quite good. The differences may be due partly to the construction of the spherical shell, which consists of two halves fitted together.

doi:10.1088/978-1-6817-4453-7ch15

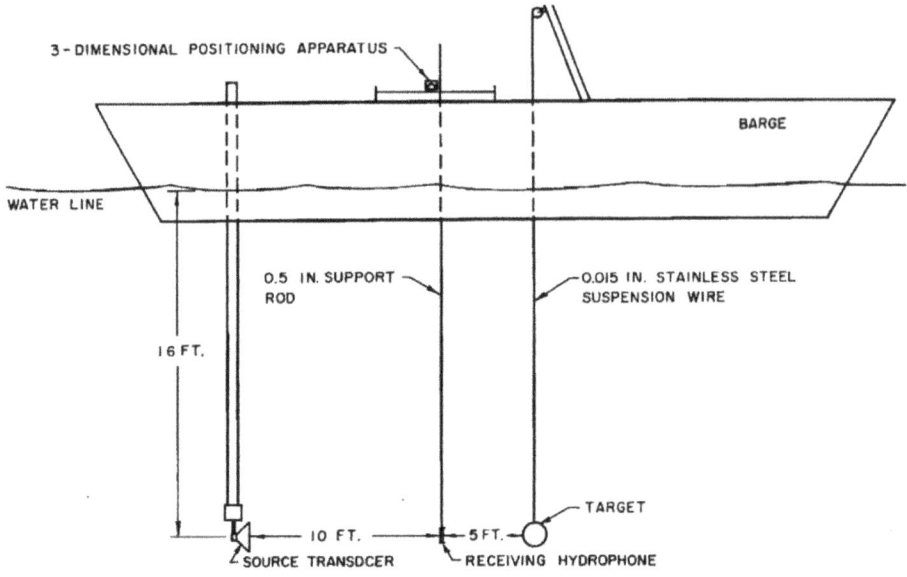

Figure 15.1. Fresh water tests under a barge.

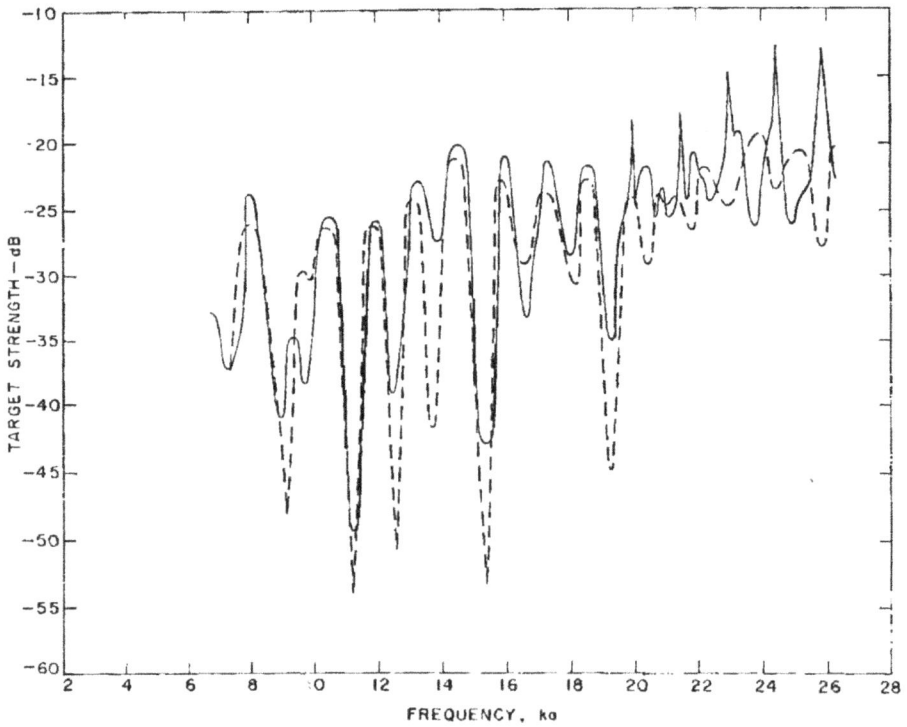

Figure 15.2. Calculated (solid) and measured (dashed) backscattered amplitudes.

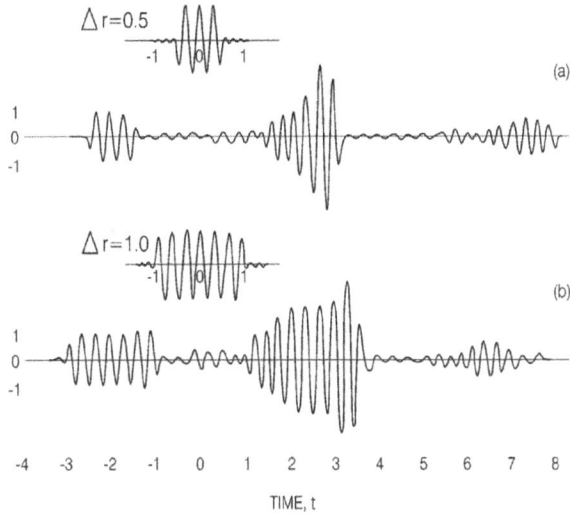

Figure 15.3. Computed results for the air-filled spherical shell.

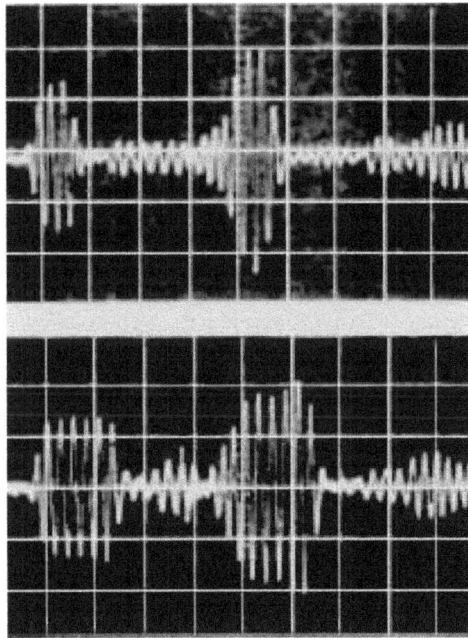

Figure 15.4. Measured data for the air-filled spherical shell.

Computed echo pulses for the air-filled spherical shell in figure 15.1 are shown below in figure 15.3 for two incident pulses with a center frequency $ka = 20$. The incident pulses are shown above the echoes.

The corresponding measured data for the air-filled shell are shown in figure 15.4. Again, it is seen that the agreement between theory and experiment is quite good.

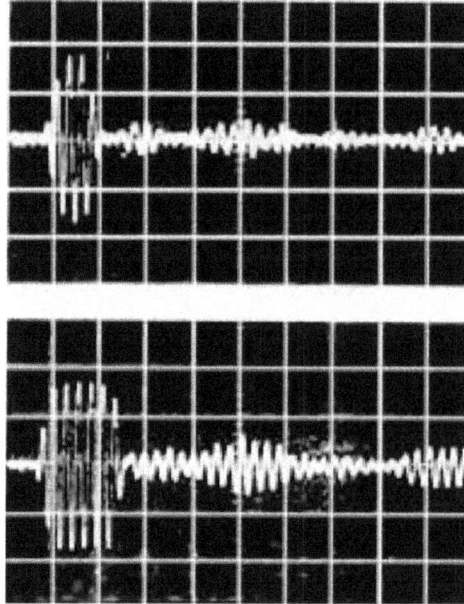

Figure 15.5. Measured data for a water-filled spherical shell.

An example shows sonar echoes can distinguish an object and its interior. Figure 15.5 shows echoes from an air-filled aluminum sphere and figure 15.5 shows echoes for the same sphere filled with water. It is seen that the first part of the echo is the same in both cases but the air-filled sphere in figure 15.5 has much stronger secondary echoes.

The secondary echoes for the water-filled sphere are weaker because sound power flows through the water in the interior. The first part of the echo looks like the incident pulse and originates from the surface of the shell closest to the source of incident sound [4].

References

[1] NASTRAN Finite-element analysis computer-aided software, one of the principal suppliers is the MSC corporation
[2] Hickling R 1958 Frequency dependence of echoes from bodies of different shape *J. Acoust. Soc. Am.* **30** 2
[3] Hickling R 1962 Analysis of echoes from a solid elastic sphere in water *J. Acoust. Soc. Am.* **34**
[4] Diercks K J and Hickling R 1967 Echoes from hollow aluminum spheres in water *J. Acoust. Soc. Am.* **41** 2

Chapter 16

Mathematical model of head impact

16.1. Description of model

Frontal head impact was modeled mathematically [1] assuming a spherical head, using data on sound transmission in brain tissue [2], together with measured flexural

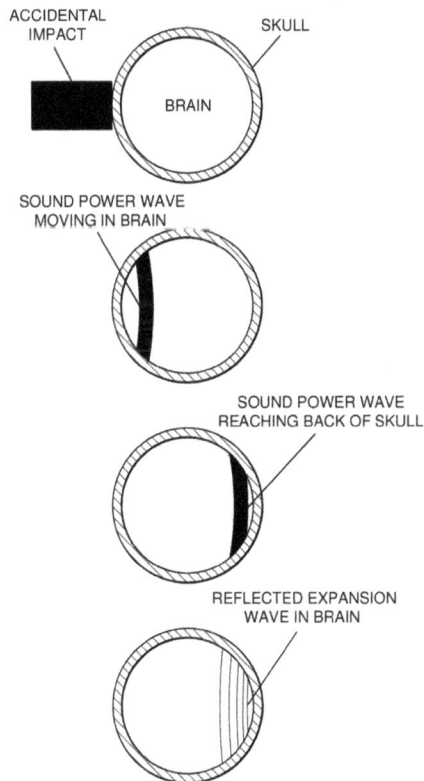

Figure 16.1. Head impact generating a pressure wave in the brain and reflected expansion waves.

bone motion to simulate the skull. Positive sound power waves transmitted through the head are reflected from the back of the head as expansion waves as shown in figure 16.1. This is known as the 'contra-coup effect', which, in severe cases, can cause the rupture of brain tissue.

References

[1] Hickling R and Wenner M L 1973 Mathematical model of a head subjected to an axisymmetric impact *J. Biomech.* **6** 115–31
[2] Ludwig G D 1950 The Velocity of sound through tissues and the acoustic impedance of tissues *J. Acoust. Soc. Am.* **22** 862–6

Chapter 17

Sound-power flow during airbag deployment in a motor vehicle compartment

17.1 Introduction

Bodily injury associated with the use of airbags has been discussed in detail, for example in [1]. Subsequently there has been a growing accumulation of medical data indicating possible injury to hearing due to airbag deployment [2–5]. The problem is aggravated in the compartments of small vehicles and by the increasing use of multiple airbag systems. Engineering solutions to reducing injury to hearing were sought some time ago at General Motors. It was found that the only viable method of reducing both the noise and overpressure of airbags was the use of a compartmentalized bag, called the breathing airbag. In this bag the outer structure inflates with gas from storage while the inner structure sucks in air from the passenger compartment through one-way cloth valves. Noise is reduced because less gas is used to inflate the bag and overpressure is reduced because at least 50% of the volume of the bag consists of air intake from the passenger compartment. Sled tests and tests in a passenger car were performed that indicate the effectiveness of the breathing airbag. Noise and overpressure were reduced by 6 dB, thus significantly reducing the risk of injury to the hearing system. In addition, use of the breathing airbag can reduce the risk of injury to an out-of-position occupant. The principle of the breathing airbag can be applied to any type of bag and is described in the following.

Overpressure is caused by the expansion of an airbag into a passenger compartment, while noise is caused by the inflation of the bag. Figure 17.1 presents a typical sound-pressure impulse due to airbag deployment that shows high frequency noise riding on low frequency overpressure.

Medical data showing possible hearing injury resulting from airbag deployment is contained in [2, 3].

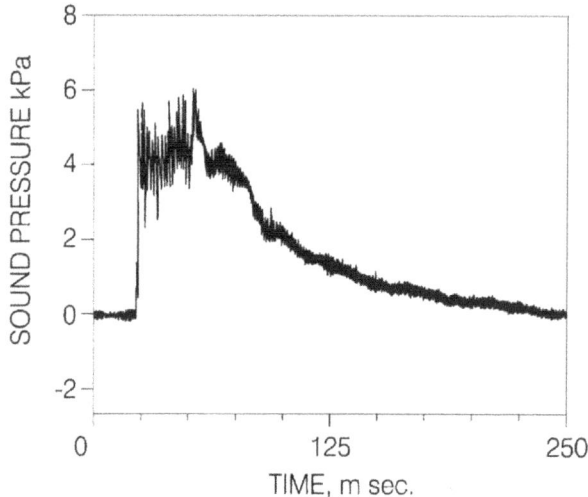

Figure 17.1. Overpressure pulse with noise superimposed.

The original GM work on reducing the noise and overpressure of airbags is described in [1]. At the time the prevailing opinion was that the risk of injury to hearing was negligible, particularly with just a driver's bag. However it was later recognized that the risk of injury to hearing was real [2, 3] and merited an investigation to find methods of reducing noise and overpressure.

The first set of tests investigated the use of silencers between the gas supply and the airbag diffuser, as shown in figure 17.2.

17.2. The breathing airbag

It was found that although several silencing devices were effective in reducing noise, they tended to increase the inflation time of the bag excessively. The best solution to reducing both noise and overpressure was found to be a compartmentalized bag, called the breathing airbag. The way this bag functions is shown in figure 17.3.

As shown, the outer side-by-side tubular structure of the bag inflates with gas from storage creating suction in the interior cavity, which draws in ambient air through one-way cloth valves. This air intake is then retained in the bag to act as part of the restraint.

The series of stills in figure 17.4 shows the breathing airbag in a sled test, with a pendulum simulating impact by a vehicle occupant, when the sled impacts the end of the run.

The impact of the sled at the end of its run deploys the bag and releases the pendulum to contact the bag. Air is sucked into the interior cavity of the bag through one-way cloth valves until the pendulum impacts the bag, at which point the valves close, retaining the air intake as part of the restraint. The test demonstrates that the bag deploys in a timely fashion and that the gas from the inflator in the outer structure combined with the air intake in the inner cavity provides an effective restraint. About 50% of the bag volume consists of air intake.

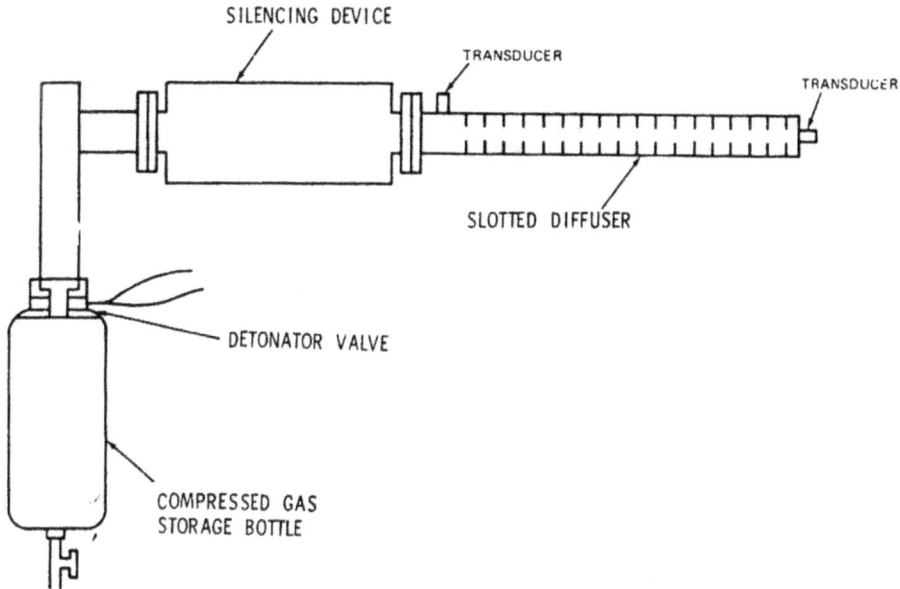

Figure 17.2. Test of silencers for reducing airbag noise.

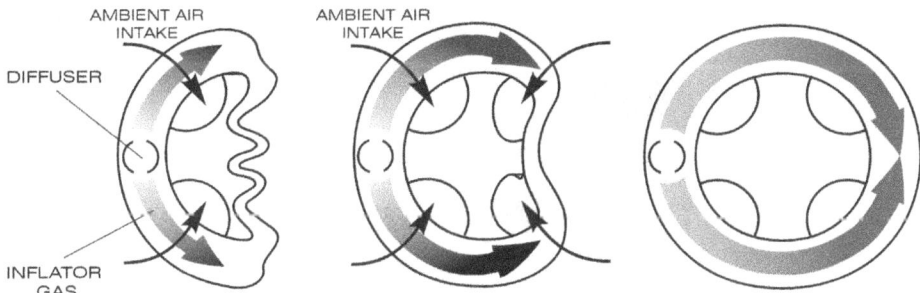

Figure 17.3. Functioning of breathing airbag.

Tests were also conducted in a passenger compartment with a breathing airbag for a front-seat passenger, shown in figure 17.5.

The pressure impulse due to deployment of this bag was compared with that of a standard bag of the same size and shape. The comparison is shown in figure 17.6.

It is seen that both the noise and overpressure are reduced by about 6 dB for the breathing airbag, which significantly reduces the risk of injury to hearing.

The concept of the breathing airbag was temporarily obscured by another type of aspirating system [5], shown in figure 17.7. Here, the operating principle is jet entrainment.

Ambient air entering through flapper valves is entrained by a gas jet from storage, flowing from the diffuser and creating an air-gas mixture in the interior of the bag. The latest data for this type of aspiration [5] indicates a percentage air intake of

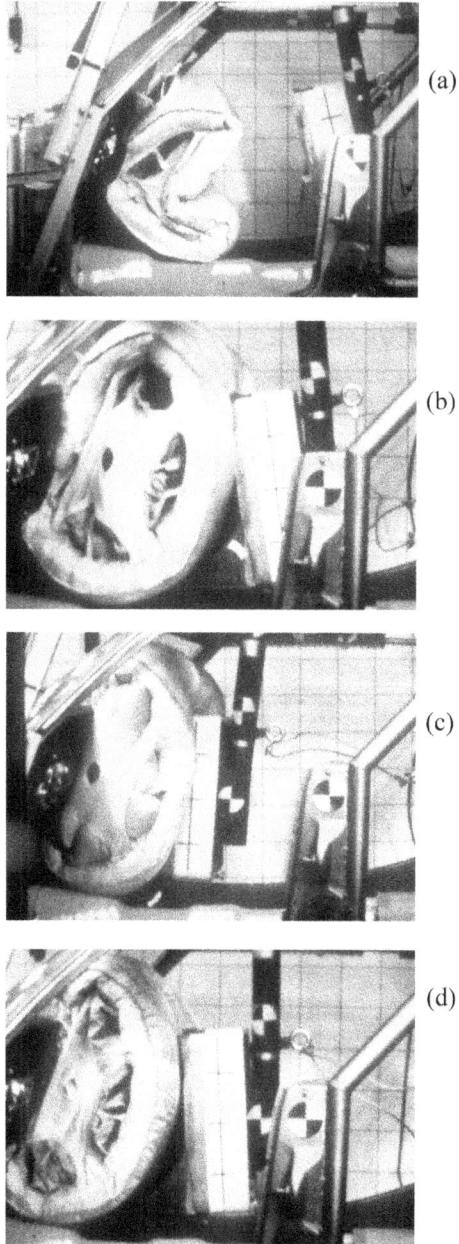

Figure 17.4. Sled test of breathing airbag.

7.8%, which is significantly less than the 50% or more obtained with the breathing airbag. It would therefore seem that the breathing airbag is superior.

An additional advantage of the breathing airbag is seen from the nature of the deployment in figure 17.4. The bag does not thrust directly at the vehicle occupant, but instead presents a relatively benign front. The breathing airbag therefore

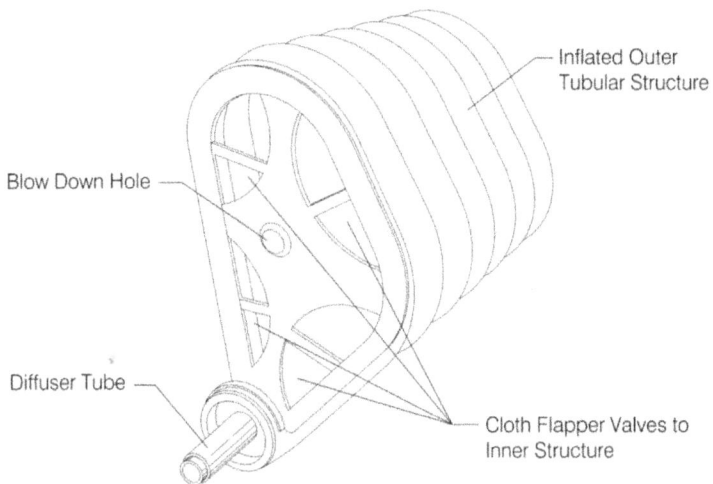

Figure 17.5. Front-seat passenger airbag.

Standard Bag

Breathing Air Bag

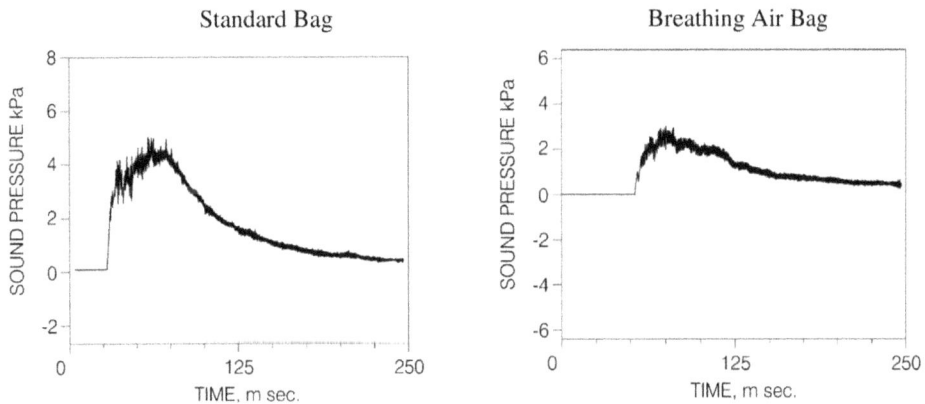

Figure 17.6. Pressure impulse due to deployment.

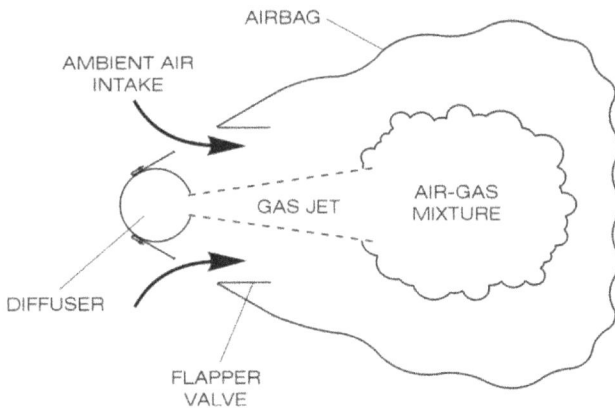

Figure 17.7. Principle of jet entrainment.

represents a form of depowering of the airbag, without an accompanying loss of restraint.

References

[1] Kent R (ed) 2003 *Air Bag Development and Performance* International Society of Automotive Engineers, 400 Commonwealth Drive, Warrendale, P15096 860
[2] Klask J 2001 Injuries of ear, nose and throat caused by airbag deployment *Laryngo-Rhino-Otologie* **80** 146–51
[3] Yaremchuk K and Dobie R A 2000 The Otologic effects of airbag deployment *J. Occup. Hear. Loss* **2** 67–73 Airbag Units, SAE paper No. 1999-01-0436
[4] Hickling R 1976 The noise of the automotive safety air cushion *Noise Control Eng.* **6** 110–21
[5] Green P W, Yu C S, Butler P B, Chen L D, Lee Y G and Wang J T Experimental analysis of aspirating airbags

Chapter 18

Uses of high-frequency, focused, pulse-echo ultrasonic probes

18.1 Description of focused, ultrasonic, pulse-echo probes

A high frequency (0.5–2 MHz), focused, pulse-echo, ultrasonic probe is depicted in figure 18.1. This shows the focused beam.

The internal structure of the transducer is shown in figure 18.2. It can be made in various sizes, depending on the application, as shown in figure 18.3.

The complete system is described in detail in [1].

18.2 Advantages

The advantages of the sound-power flow of the pulse echo ultrasonic probes are as follows:

- Background noise interference is virtually eliminated because only the reflected signal can reach the transducer with sufficient strength to be measured.

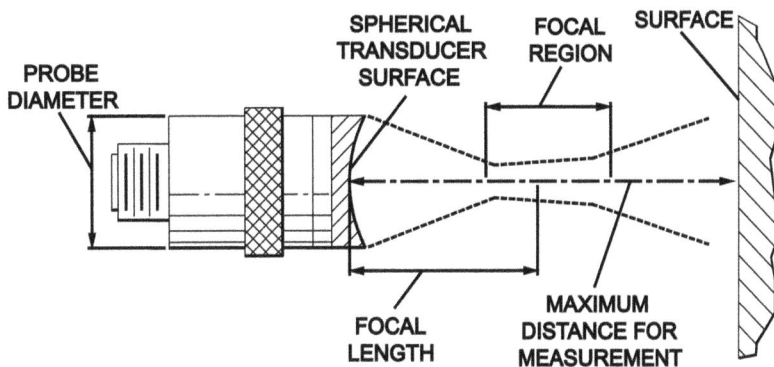

Figure 18.1. Focused, high frequency, ultrasonic, pulse-echo probe.

doi:10.1088/978-1-6817-4453-7ch18

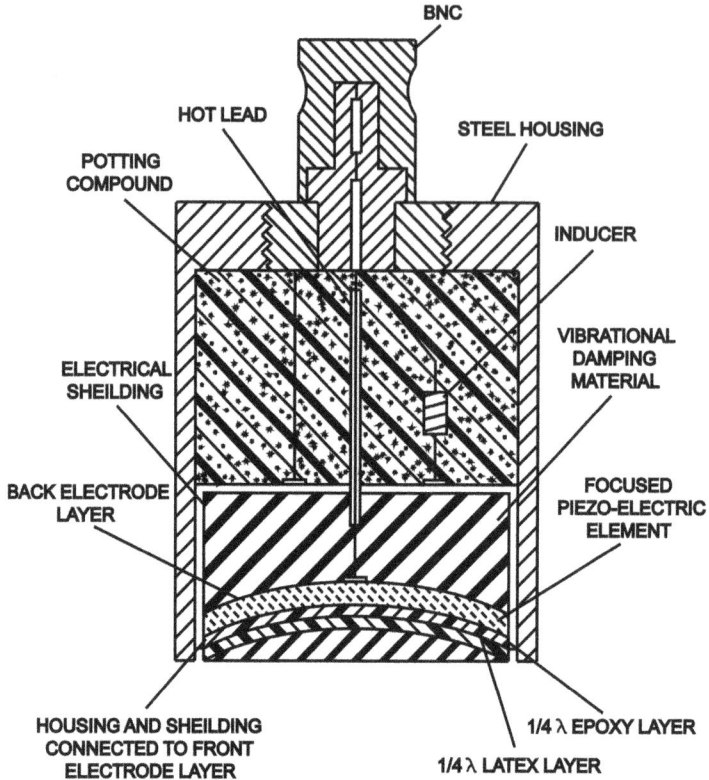

Figure 18.2. Internal structure of a transducer.

Figure 18.3. Different sizes of transducer.

- High frequency ultrasound from more distant sources is absorbed by air.
- Probe can easily be used in liquids as well as air.
- Focusing to a spot size, typically about 1 mm in diameter, provides good lateral resolution.
- In contrast to optical reflectivity, ultrasonic reflectivity remains almost constant for most solid and liquid surfaces.

- A measurable echo is obtained even from soft surfaces such as carpeting and fiberglass.
- Ultrasonic gauging systems are inherently less costly, smaller, lighter and more rugged than optical and other non-contact gauging devices.
- High-frequency ultrasound is well above frequencies that can affect human hearing.

18.3 Vibrometry

Figure 18.4 shows an ultrasonic probe measuring the vibrations at the end of the shaft of a small electric motor.

18.4 Condition monitoring

Figure 18.5 shows two probes monitoring the condition a large steam turbine shaft, such as is used in power generation or in ships.

A photograph of an actual shaft is shown in figure 18.6 indicating its typical size.

18.5 Feedback control of a robot arm

Tests have shown that a focused probe can guide a robot arm over a surface and determine the shape of the surface, as shown in figure 18.7.

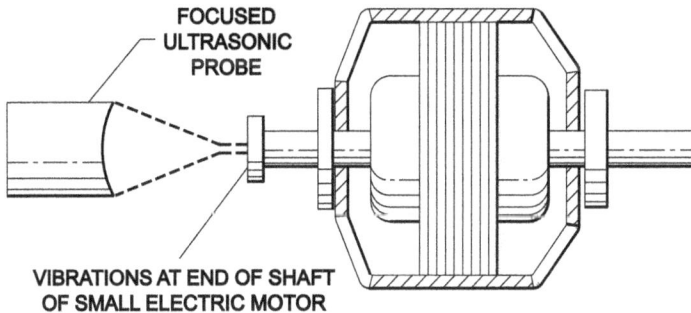

Figure 18.4. Measuring vibrations at the end of a shaft.

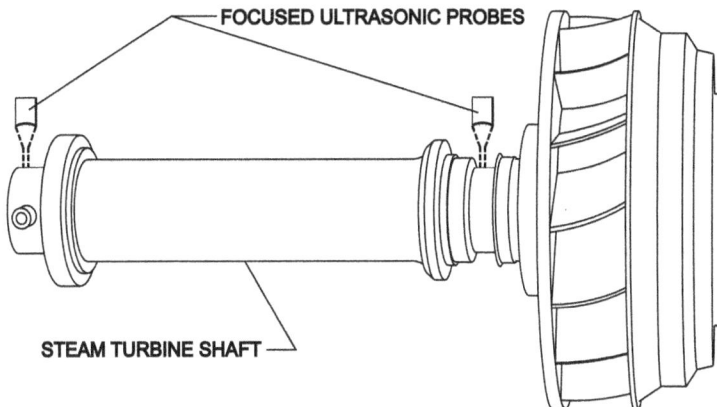

Figure 18.5. Condition monitoring of a steam turbine shaft.

Figure 18.6. Turbine shaft.

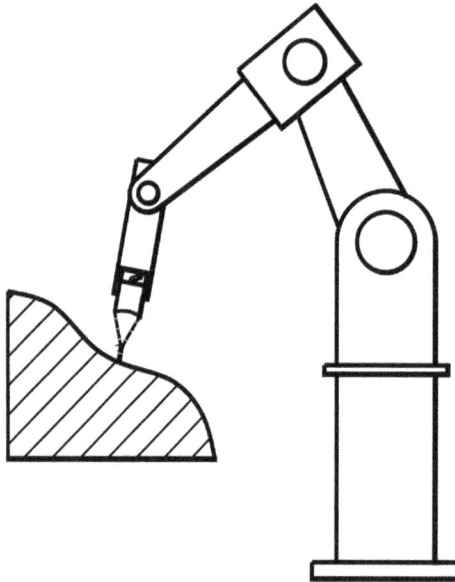

Figure 18.7. Focused probe guiding a robot arm.

18.6 Detecting missing and out-of-position locking keys in engine valves

In the manufacture of engines on a production line, it is necessary to monitor the locking keys of engine valves to determine whether they are missing or out of position. A focused probe with a very small spot size can be used to inspect the upper surface of an engine valve as it passes underneath the probe held in a fixed position. Figure 18.8 depicts the process. The probe is fixed and the valve is moving to the right.

The process was demonstrated in laboratory tests with an engine head on a short conveyer belt as shown in figure 18.9. Figure 18.10 shows an engine valve with a raised out-of-position locking key.

To demonstrate the accuracy of the ultrasonic probe, the probe was scanned back and forth over the engine valve in small steps. The result is shown in figure 18.11. Even the slightly raised letter H on the rim of the valve, is clearly discernable. In general the probe can be used to measure surface roughness.

18.7 Measuring the acid level in battery compartments

Another use of a focused probe on a production line is to inspect the acid level in batteries. Here the probe measures the distance to the acid level surface as a battery passes underneath, as shown in figure 18.12. The battery is moving from left to right and the probe is fixed.

Measuring the acid level was demonstrated in laboratory tests as shown in figure 18.13. The acid levels are shown on the screen directly behind the battery, which is moving on the short conveyer belt. Comparison with dipping tests showed excellent agreement. The air motion due to the relatively slow motion of the battery on the

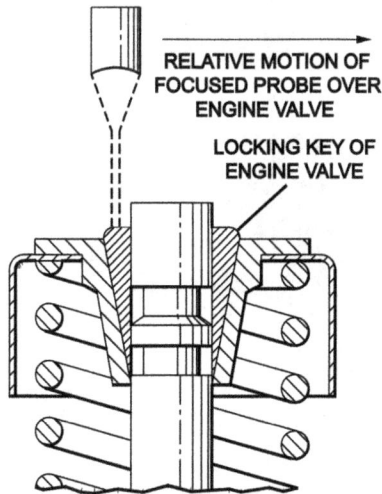

RELATIVE MOTION OF FOCUSED PROBE OVER ENGINE VALVE

LOCKING KEY OF ENGINE VALVE

Figure 18.8. Focused probe inspecting the upper surface of an engine valve.

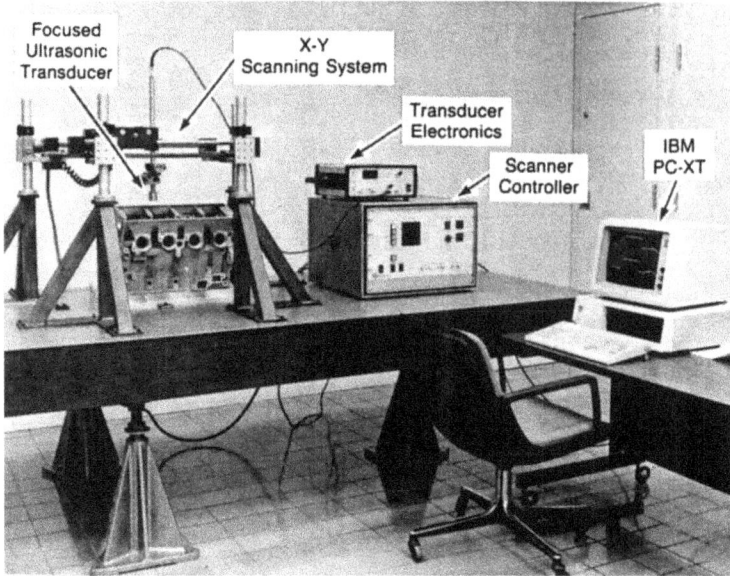

Figure 18.9. Scanning engine heads on a conveyer belt.

Figure 18.10. Engine valve with raised out-of-pssition locking key.

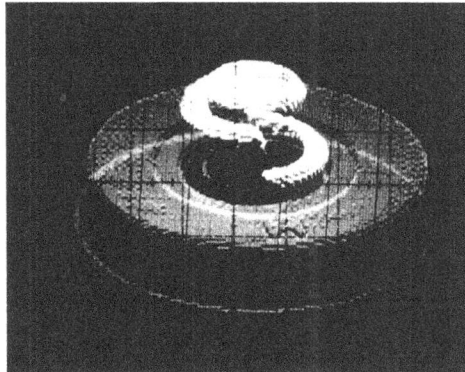

Figure 18.11. Accuracy of multiple scans of an ultrasonic probe.

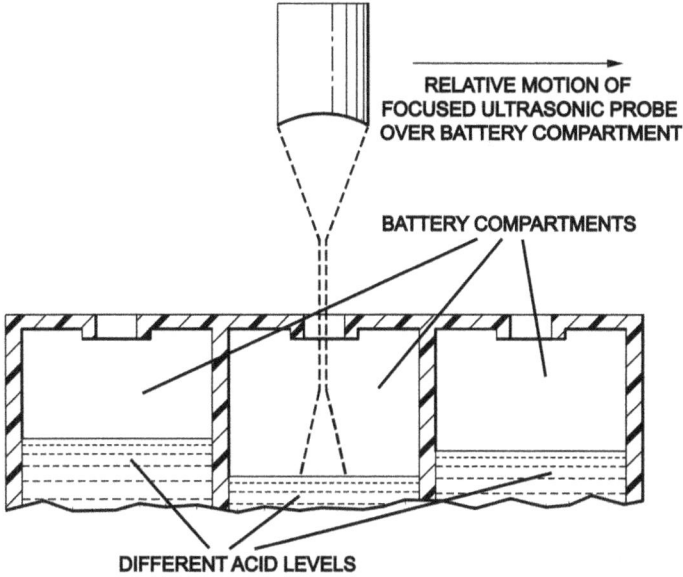

Figure 18.12. Focused probe measuring the acid levels in a battery.

Figure 18.13. Measuring the acid level in a battery.

conveyer belt does not affect the ultrasonic beam and the accuracy of the measurements.

18.8 Gauging

Focused probes can also be used to monitor and control items during manufacturing, as shown below in figures 18.14 through 18.16.

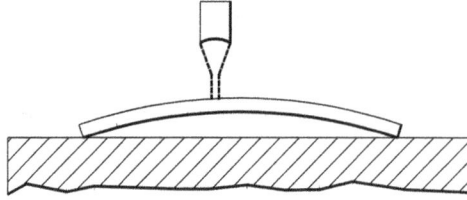

Figure 18.14. Warpage of a plate detected by a probe.

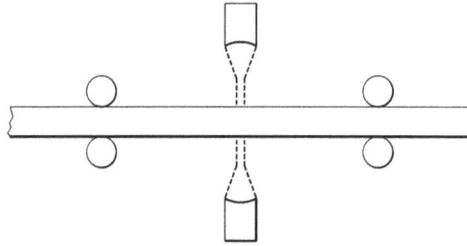

Figure 18.15. Pair of probes monitoring the thickness of a plate on a production line.

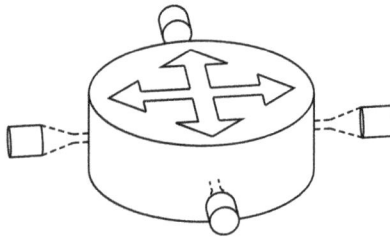

Figure 18.16. Positioning circular piece with four focused probes.

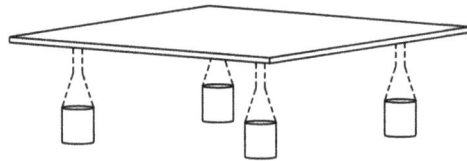

Figure 18.17. Four focused probes monitoring horizontal positioning of a plate.

18.9 Testing the effectiveness of welds

Weld testing with a focused non-contact probe is shown in figure 18.18.

18.10 Detection of buried objects with an array of non-contact probes

The system is shown in figure 18.19 and is described in [2].

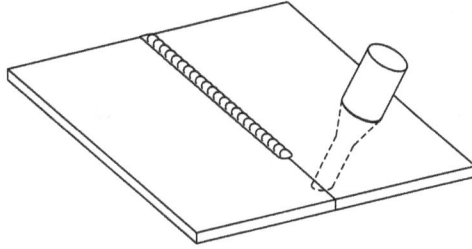

Figure 18.18. Weld testing.

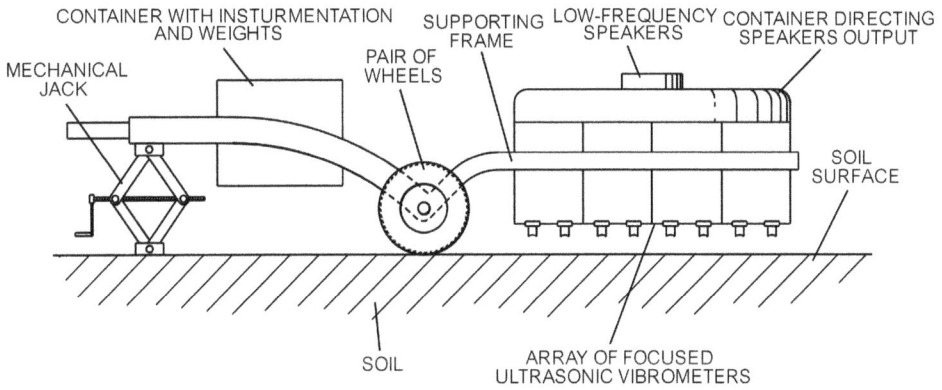

Figure 18.19. System for detecting buried objects.

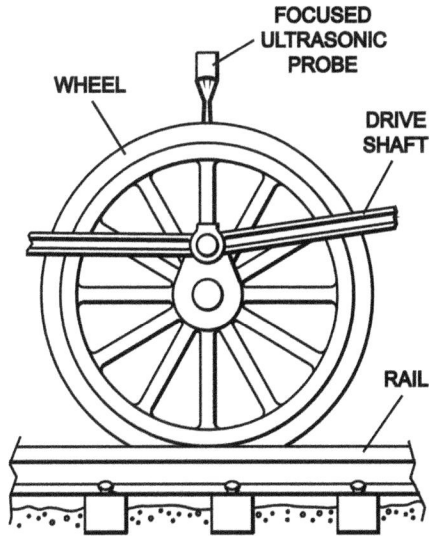

Figure 18.20. System for detecting railroad problems.

The system can be made in various sizes depending on the application and the terrain. It can be used, in particular, for detecting non-metallic land mines.

18.11 Detecting flaws and problems in a railroad and a train wheel

The system is shown in figure 18.20 and described in [1]. It can be used for a number of purposes including testing for excessive cornering speed.

References

[1] Hickling R 2012 Non-contact focused ultrasonic probes for vibrometry, gauging, condition monitoring and feedback control of robots US Patent, No. 8,296,084

[2] Hickling R 2010 Detection of buried objects using an array of non-contact ultrasonic vibrometers US Patent No. 7,751,281

Chapter 19

Detecting insect pests in agricultural commodities

19.1 Detecting insects in cotton bolls

It is necessary to discover infestations of insects in cotton fields before harvesting so that insecticides can be applied appropriately and quarantine used where necessary. Infestations have to be found at an early stage, by sampling bolls gathered from the fields, according to their location, and cutting the bolls open to detect insects, principally pink bollworm. This tedious labor is performed principally by summer students, with the risk of cutting fingers and hands, and of not detecting every insect. Insects make sounds in the bolls and this can be detected by locating bolls individually on an array of acoustic sensors, in a soundproof box and listening, one at a time, to each sensor [1, 2]. An example from [1] is shown in figure 19.1. The listening can be done either by a person or with an automated system. It can also be done simultaneously at all the sensors, and possibly using robots to locate the bolls on the sensors.

A sensor is shown in figure 19.2. It consists of a stethoscope head and can include a small fluid-filled bag for improving sound transmission to the head, as shown in figure 19.2(b).

The system can also be used for other items such as eggs and cherries.

19.2 Detecting insects in stored grain

In the United States, about 10%, or a billion dollars worth of stored grain, is destroyed annually by insects. Grain loss is much higher in some developing countries [3]. Insects also affect the quality of grain exports [4]. Early detection is obviously important. A number of detection methods have been investigated [5, 6]. For automated systems in bulk-stored grain, the most practical method appears to be acoustic detection [6], using microphones designed to penetrate the grain as shown in figure 19.3.

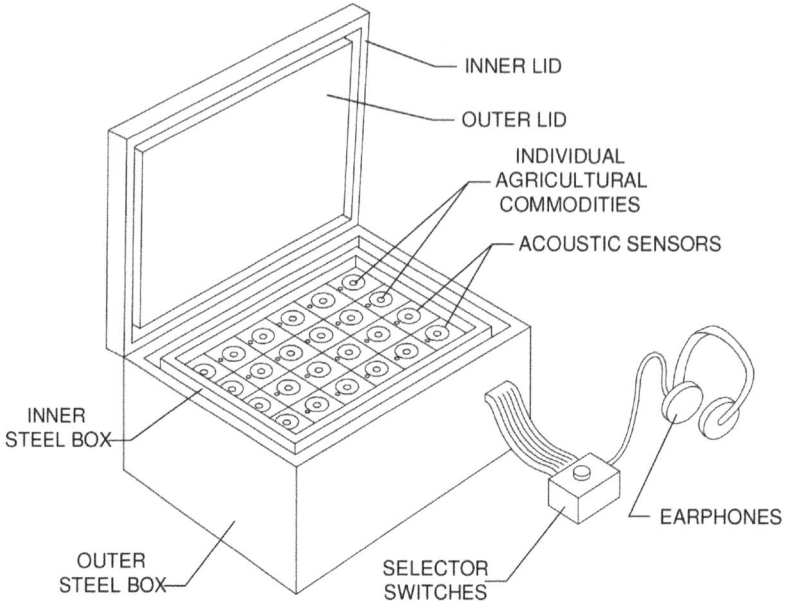

Figure 19.1. Array of acoustic sensors in a soundproof box.

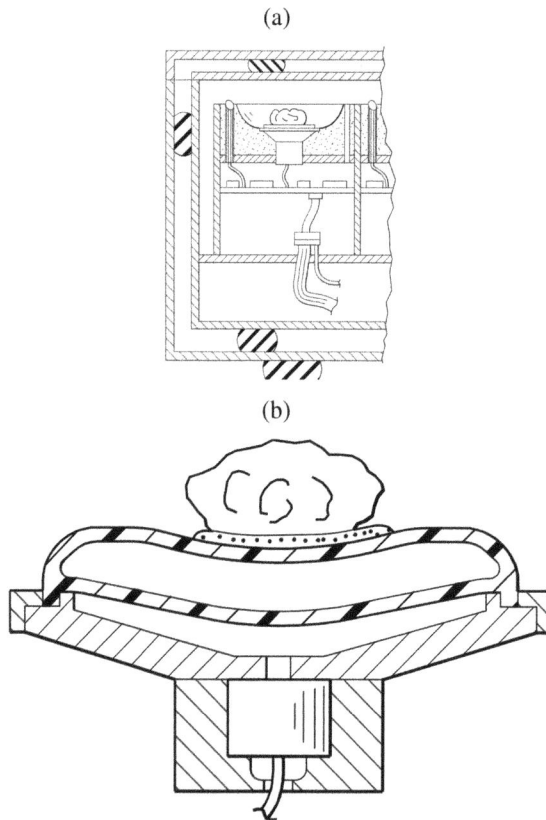

(a)

(b)

Figure 19.2. Sensor arrangement.

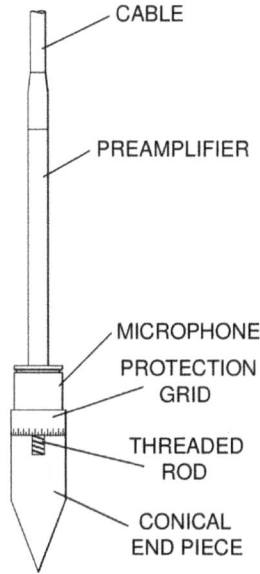

Figure 19.3. Microphone system for grain.

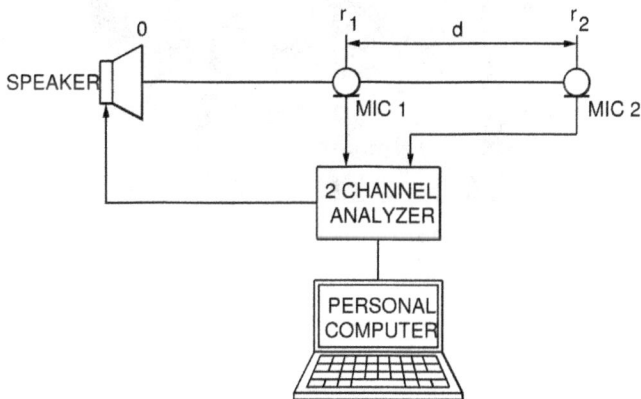

Figure 19.4. System for measuring transmission in grain.

The system for measuring the acoustic transmission through grain between two microphones a known distance d apart, is shown in figure 19.4.

The transmission occurs mainly through the passageways between grains, as shown in figure 19.5 for grain saturated with different gases.

Various kinds of stored grain were tested [6]. These are shown in the photograph, figure 19.6. They are, from left to right: soft wheat, sorghum, brown rice, soybeans, corn, and hard wheat.

The passageways between grains depend on the type of grain, as shown in figure 19.7.

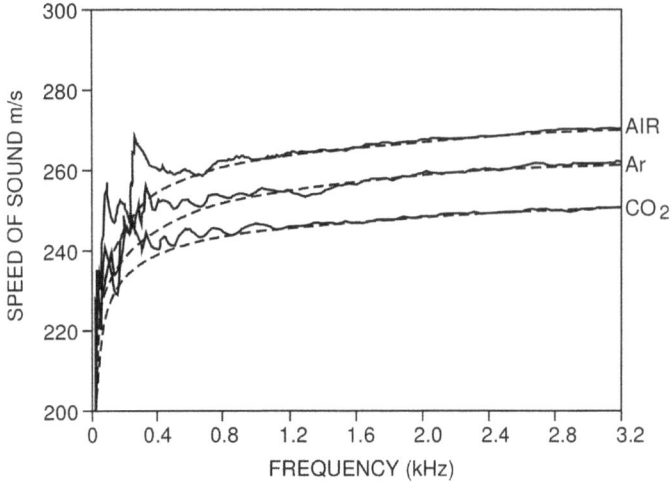

Figure 19.5. Transmission speed in grain saturated with different gases.

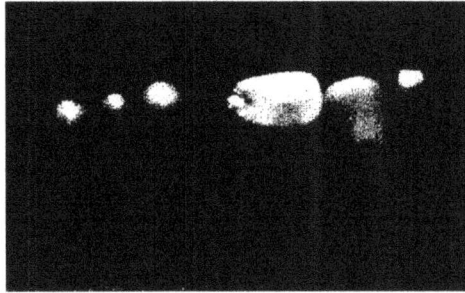

Figure 19.6. Various kinds of stored grain.

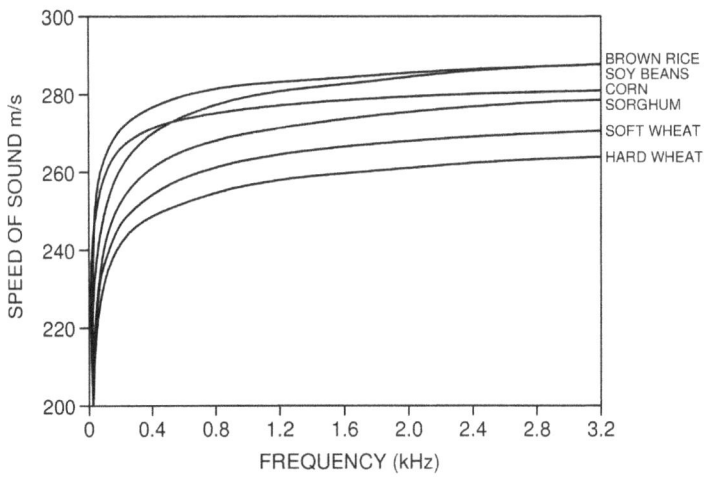

Figure 19.7. Speed of sound in various kinds of grain.

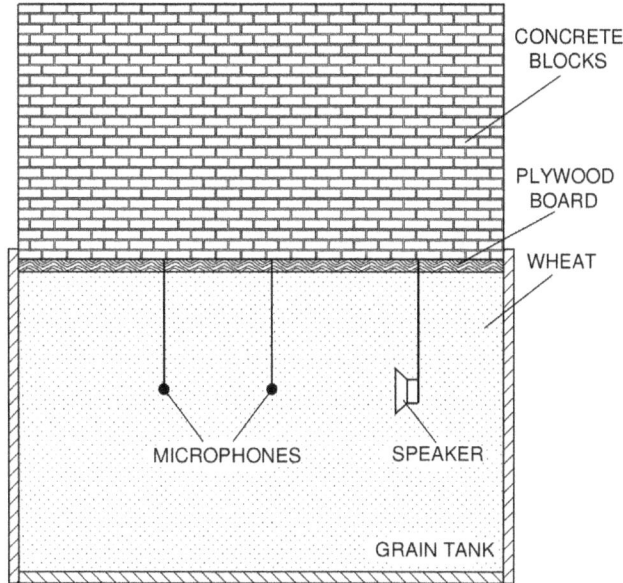

Figure 19.8. Testing the effect of depth of grain.

It was shown that the depth of the grain does not affect the sound transmission significantly. The tests were conducted using the apparatus shown in figure 19.8, with the pressure of concrete blocks simulating the depth of grain.

References

[1] Hickling R, Lee P, Wei W and Chang S T 1997 Acoustic sensor system for insect detection *US Patent No.* 5,616,845

[2] Hickling R *et al* 1994 Multiple acoustic sensor system for detecting pink bollworm in bolls *Proceedings of National Beltwide Cotton Conferences (San Diego, CA)*

[3] Food and Agriculture Organization of the United Nations 1977 Analysis of FAO Survey of Post-Harvest Crop Losses in Developing Countries (Rome, Italy)

[4] Leesch J G, Arthur F H and Davis R 1990 Three methods of aluminum phosphide application for the in-transit fumigation of grain deep-draught bulk cargo ships *J. Econ. Entomol.* **83** 1459–67

[5] 1991 United States Department of Agriculture, maintaining insect-free farm-stored grain, agricultural information bulletin No. 580

[6] Hickling R, Wei W and Hagstrum D W 1990 Studies of sound transmission in various types of stored grain for acoustic detection of insects, (note error in figure 1 of paper. It should include hard wheat at end of sequence of six grains) *Appl. Acoust.* **50** 263–78

Sound-Power Flow
A practitioner's handbook for sound intensity
Robert Hickling

Chapter 20

Holography of liquid droplets

20.1 Introduction

The purpose in this chapter is to determine the size of individual liquid droplets in a small cloud, generated by ultrasonic sound-power flow [1]. The size of a droplet cannot be determined accurately from a photograph, such as in figure 20.1

However there is a relation between the size of droplet and the angular positions of the peaks in the geometric form of its scattered light pattern, computed using classical Mie theory. Hence if a suitable portion of the radiation pattern is known, it can be used to determine droplet size with much greater accuracy. Holography was used to determine a portion of the radiation pattern of droplets in a bubble cloud. The holograms were made using a single-mode, Q-switched ruby laser, as described by Siebert [2].

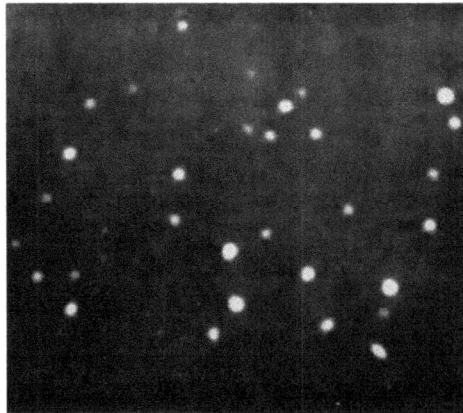

Figure 20.1. Photograph of a bubble cloud.

doi:10.1088/978-1-6817-4453-7ch20

20.2 Generating liquid droplets using ultrasonic sound-power flow

A cloud of liquid droplets was generated [1] with an ultrasonic driver, as shown in figure 20.2.

20.3 Use of holography

A series of scattered peaks for a particular droplet was then determined using front lighting R from a ruby laser [2] onto a holographic plate H, as shown in figure 20.3.

The scattered peaks of a droplet on the hologram H was determined on a TV, using a helium-neon laser L contiously scanning the reverse side of the hologram, as in figure 20.4.

These peaks can then be compared with peaks computed using the Mie scattering equations for a droplet size, chosen to fit the measured peaks. An example of such a comparison is shown in figure 20.5, where the ragged peaks were measured and the smooth peaks were computed.

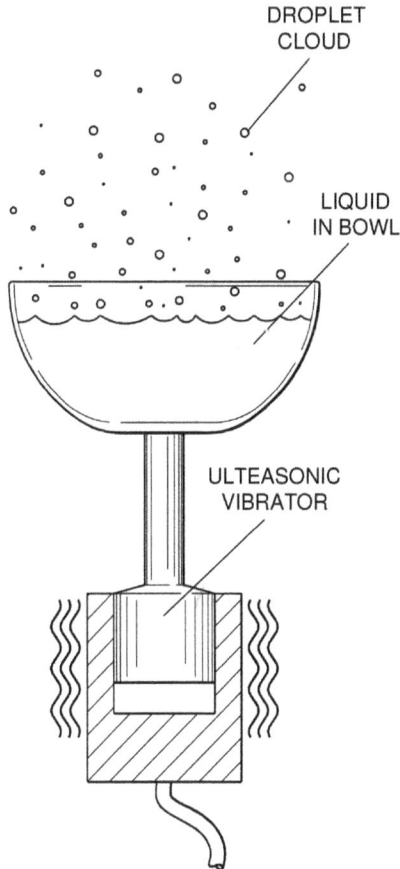

Figure 20.2. System for generating a cloud of droplets.

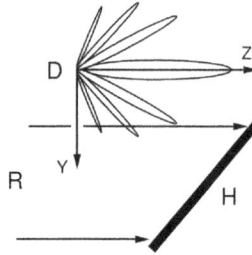

Figure 20.3. Recording scattered peaks on a hologram.

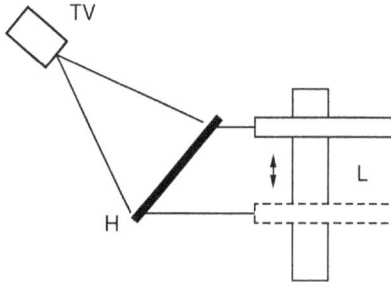

Figure 20.4. Helium-neon laser scanning hologram.

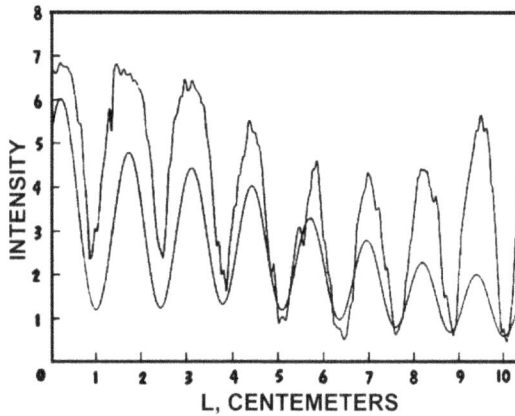

Figure 20.5. Comparison between measured and computed scattered peaks.

The fit btween the peaks then determines the size of the particular droplet in the ultrasonically generated cloud.

20.4 Computer-generated holograms

Computer-generated holograms were first developed by Meyer and Hickling [3] and applied by Hickling to sperical liquid droplets [4]. They have since been used by many others but their use has still to be developed. Possibly they could be used by illustrators at Disney Studios.

References

[1] Hickling R 1969 Holography of liquid droplets *J. Opt. Soc.* **59** 1334–9
[2] Siebert L D 1967 Front-Lighted laser holography *Appl. Phys. Lett.* **11** 326–9
[3] Meyer A J and Hickling R 1967 Holograms synthesized on a computer-operated cathode-ray tube *J. Opt. Soc. Am.* **57** 11
[4] Hickling R 1968 Scattering of light by spherical liquid droplets using computer-synthesized holograms *J. Opt. Soc. Am.* **58** 4

Chapter 21

Wooden sounding boards

21.1 Introduction

Wooden sounding boards have a number of uses, such as behind pulpits in large older churches and with various types of musical instruments including pianos, organs, violins, guitars, and harps. They can also be used in large classrooms. Use of the term 'sounding board' can describe a person, on whom one tries out ideas. Sounding boards are often replaced using microphones and loudspeakers.

21.2 Acoustic characteristics of wood

The acoustic characteristics of some kinds of wood are described in [1]. These depend on the direction of the wood fibers, as shown in table 21.1.

There does not appear to be data on the speed of sound in spruce, probably because there are many varieties of spruce. The Wright Brothers used spruce in their wing and propeller structures. [2]

Table 21.1. Acoustical characteristics of wood.

Ash, along the fiber	4670 m s^{-1}
across the fiber	1390 m s^{-1}
along the rings	1260 m s^{-1}
Beech, along the fiber	3340 m s^{-1}
Elm, along the fiber	4120 m s^{-1}
Maple, along the fiber	4110 m s^{-1}
Oak, along the fiber	3850 m s^{-1}
Spruce	?

21.3 Pulpits and large classrooms

A pulpit with a sounding board behind is shown in figure 21.1.

In large classrooms, the black or writing board can act as a sounding board.

21.4 Musical instruments

A grand piano is shown in figure 21.2. The lid is raised to release the sound reflected by the sounding board at its base. This is often made of spruce. The spruce boards are glued together along the grain. Spruce's has a high strength to weight ratio offering strength sufficient to withstand the downward force of the strings. The best piano makers use quarter-sawn, defect-free spruce of close annular grain, carefully seasoning it over a long period before fabricating the soundboards. This is the material that is used in quality acoustic guitar soundboards. Inexpensive pianos often have plywood sounding boards.

A violin is shown in figure 21.3. Typically the top is made of two pieces of fine grain Italian spruce and the sounding board in the back is made of slab cut Oregon maple.

21.5 Amphitheaters, cathedrals, mosques, and stadia

In early times, before the use of electronics, considerable use was made of stone structures to reflect and augment sound for gatherings of people. Amphitheaters were tiered, mostly semicircular and sometimes oval in shape. An example of a semicircular shape is shown in figure 21.4.

There were many semicircular amphitheaters in Roman and Greek territories. A well-known oval amphitheater is the Coliseum in Rome. Amphitheaters for popular gatherings were unknown in ancient Egypt and China, presumably because popular

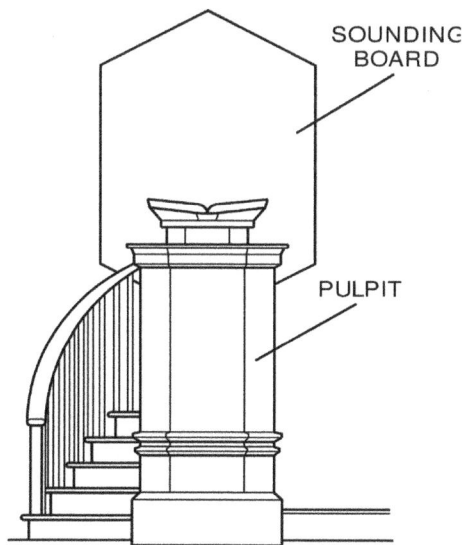

Figure 21.1. Pulpit with sounding board

Figure 21.2. Grand piano.

Figure 21.3. Violin.

gatherings were not encouraged. In the Christian world, mainly in Europe, cathedrals were constructed to protect congregations and priests from inclement weather and to provide significant, long-distance landmarks. Their interiors augmented the sound of priests, choirs, congregations and church bells. Open pagan structures such as Stonehenge in Southern England also provided gathering places. These and the pyramids in Egypt appear to exist mainly for astronomical purposes.

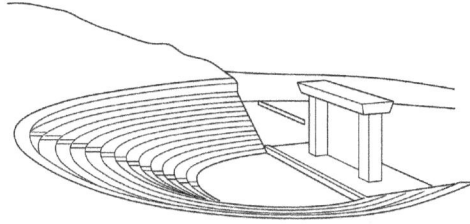

Figure 21.4. Semicircular amphitheater.

Mosques, which can be significant in size and beauty, provide places of worship. The call to prayer from mosque minarets is now generally broadcast electronically. A central focal place for Moslems is Mecca in Saudi Arabia, which is very large and open to the air. Muslims, who are able, are required to make a pilgrimage there at least once during their lifetime.

Other large open stone structures that augment sound occur in old Mayan communities.

References

[1] CRC Handbook of Chemistry and Physics 2009 90th edn ed R David and Lide
[2] McCullough D 2016 The Wright Brothers p 336 (New York: Simon and Schuster)

Chapter 22

Use of deturbulators to improve performance of wind farms

22.1 Introduction

Sinha [1], aided by James E Hendrix, developed the use of deturbulators to improve the lift of airfoils, specifically for the wings of gliders and aircraft. Deturbulators were also applied to heavy trucks. The purpose in this chapter is to indicate how deturbulators may be used to increase the electrical output and reduce the noise of wind turbines in wind farms.

22.2 Wind turbines

Wind turbines fall into two basic groups: the horizontal-axis variety, as shown on the left in figure 22.1, and the vertical-axis design, like the eggbeater-style Darrieus model on the right, named after its French inventor [2]. Horizontal-axis wind turbines typically either have two or three blades. These three-bladed wind turbines are operated 'upwind,' with the blades facing into the wind. The Darrieus type is independent of wind direction. Paraschivoiu's book [3] provides much greater detail. Examples of the two types are shown in figure 22.2.

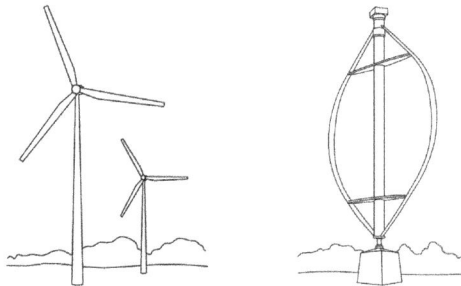

Figure 22.1. Two types of wind turbine.

doi:10.1088/978-1-6817-4453-7ch22   22-1

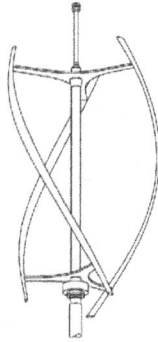

Figure 22.2. Darrieus wind turbine with twisted blades.

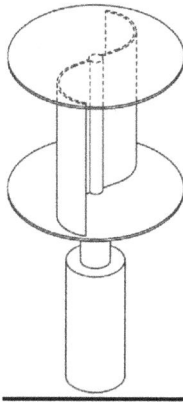

Figure 22.3. Savonius type of wind turbine.

Figure 22.4. Deturbulator applied to the leading edge of an airfoil.

The Darrieus type can have twisted blades as shown in figure 22.2, which can be used in winds of continuously shifting direction.

The Darrieus type can be combined with the Savonius type [4], an example of which is shown in figure 22.3. The main advantage of this combination is to improve efficiency at low wind speeds.

22.3 Deturbulators

Hendrix originated the term deturbulator. The original deturbulator device was for an aircraft or glider wing with a small rear-facing step near the leading edge of the airfoil (about 0.0025 inches high) and about one inch long, as shown in figure 22.4. A photograph of a glider test is shown in figure 22.5.

Figure 22.5. Glider test.

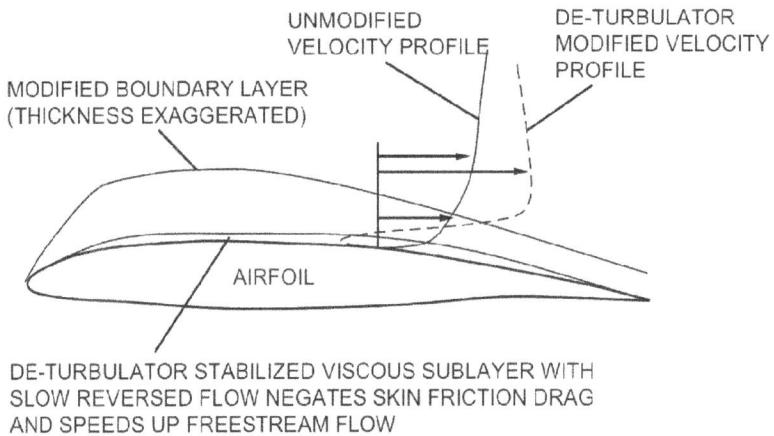

UNMODIFIED VELOCITY PROFILE

DE-TURBULATOR MODIFIED VELOCITY PROFILE

MODIFIED BOUNDARY LAYER (THICKNESS EXAGGERATED)

AIRFOIL

DE-TURBULATOR STABILIZED VISCOUS SUBLAYER WITH SLOW REVERSED FLOW NEGATES SKIN FRICTION DRAG AND SPEEDS UP FREESTREAM FLOW

Figure 22.6. Results of tests of deturbulators on an airfoil.

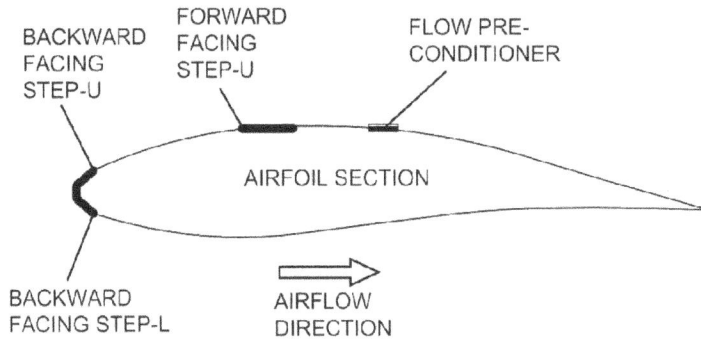

BACKWARD FACING STEP-U

FORWARD FACING STEP-U

FLOW PRE-CONDITIONER

AIRFOIL SECTION

BACKWARD FACING STEP-L

AIRFLOW DIRECTION

Figure 22.7. Additional deturbulator methods.

Results with gliders [1] are shown graphically in figure 22.6. The tests showed a noticeable improvement in glider performance.

Additional methods can be used with gliders, as shown in figure 22.7 [1].

Such methods were also used with aircraft wings and for heavy trucks.

22.4 Use of deturbulators on vertical-axis wind turbines

They can also be used for the Darrieus vertical-axis wind turbines. It will be necessary, however, to conduct tests to determine which methods are most effective for improving wind turbine performance.

References

[1] Sinha S K Method for using a flexible surface deturbulator to reduce the aerodynamic drag of bluff bodies U.S. Provisional Patent Application No. 60/784,047, Filing Date 03/20/2006; Foreign Filing License Granted 04/07/2006.

[2] Darrieus G J M 1931 Darrieous rotor *US Patent No.* 1,835,018

[3] Paraschivoiu I 2002 Wind Turbine Design with the Emphasis on Darrieus Concept p 397 (Montreal, Canada: Polytechnic International Press)

[4] Savonius S J 1928 *US Patent No.* 1,697, 574 (1925) and *US Patent No.* 1, 766,765

Chapter 23

Sound-power flow in medical ultrasonics

23.1 Medical uses of ultrasound

Different types of image can be formed using ultrasound [1]. The commonest is the image, which displays the acoustic impedance of a two-dimensional cross-section of tissue. Other types of image display the blood flow motion in tissue over time, the location of blood, the presence of specific molecules, the stiffness of tissue, and the structure of a three-dimensional region of tissue such as the liver and the womb.

Ultrasound has the following advantages:
1. Provides images in real-time.
2. Can be used at the bedside.

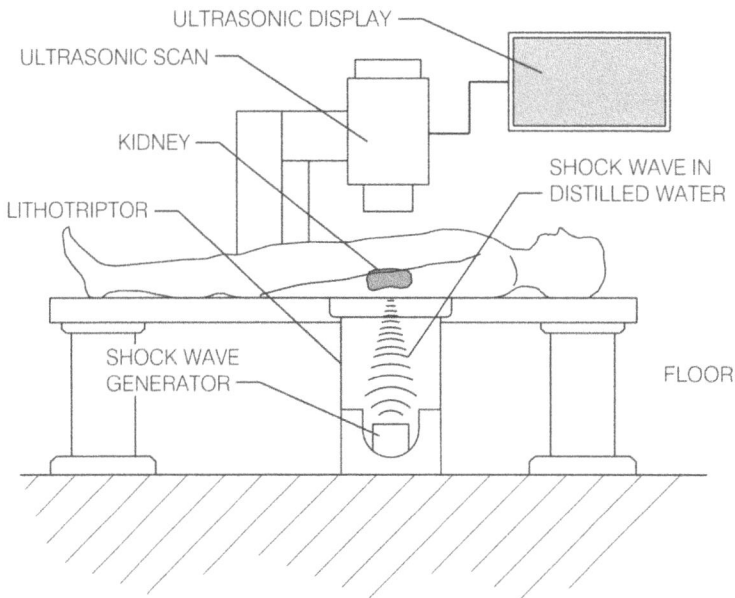

Figure 23.1. Use of a lithotripter.

3. Lower in cost.
4. Does not use harmful ionizing radiation.

The drawbacks are:
1. Difficulty imaging structures behind bone and air,
2. Need for a skilled operator.

23.2 Use of shock waves in lithotripsy

A procedure called lithotripsy uses shock waves to break up stones in the kidney or bladder or the tube that carries urine from the kidneys to the bladder. Figure 23.1 is a representation of the use of a lithotripter. Ultrasonic scanning provides an image of the progress in the breakup of the stones, which may or may not be visible to the patient. There can be a danger due to cavitation erosion.

Reference

[1] Hill C R, Bamber J C and ter Haar G R (ed) 2015 The Physical Principles of Medical Ultrasonics 2nd ed (Hoboken, NJ: Wiley)

Chapter 24

Metamaterials

24.1 Introduction

Metamaterials are concerned with the use of sound and vision. Acoustic metamaterials are artificially fabricated materials designed to control and manipulate sound waves in gases, liquids, and solids [1]. Controlling sound waves using acoustic metamaterials can be extended into the negative refraction domain. Control of the various forms of sound waves is mostly accomplished through the bulk modulus, mass density, and chirality [2]. A molecule is chiral if there is another molecule, real or potential, that is of identical composition, but which is arranged in the form of an image that cannot be superimposed, e.g. backs of a pair of hands.

24.2 Acoustic cloak

An acoustic cloak prevents objects from reflecting sound waves. This can be used to build sound-proof homes, advanced concert halls, or stealth warships. The mathematics and physics behind acoustic cloaking has been known for several years. The idea of acoustic cloaking is to deviate the sound waves so that they move around the object being cloaked. This is not straightforward. The solution is to use acoustic metamaterial. Making an acoustic metamaterial depends on the material's mass density and elastic constant. Research [3] has indicated that there are metamaterials, which simultaneously have a negative bulk modulus and mass density.

24.3 Invisibility cloak

Recent work [4] at the University of California at Berkeley has developed an invisibility cloak that uses light channeling metamaterials. These have features that are much smaller in size than the wavelength of light, which allows them physically to re-route the light waves coming in. They delay the light and delay the reflection, in such a way that every point reflects light as if from a flat surface, like a mirror. For now, it only works if the person wearing the cloak remains still. If the person moves, the person is easy to see because movement disrupts the metamaterial's ability to re-route the light.

References

[1] Wagnière G H 2007 On chirality and the universal asymmetry: reflections on image and mirror image (New York: Wiley-VCH) p 241

[2] Crastar R V and Gnenneau S (ed) 2013 Editors and Authors Acoustic Metamaterials: Negative- Refraction, Imaging, Lensing and Cloaking (Dordrecht, Germany: Springer Series in Materials Science) p 320

[3] Zhou X and Hu G 2007 Acoustic Metamaterials and Wave ControlFrontier Research in Computation and Mechanics of Materials)

[4] Zhang X 2015 (Berkeley, CA: University of California Press)

Chapter 25

Cosmic sound waves

As indicated in the Big Bang theory [1], there was an ultra high-density region of primordial plasma at the origin of space-time. This high-density region had gravitationally attracted matter in it, while at the same time the heat of matter–photon interactions created a significant outward pressure. The counteracting forces of gravity and pressure generated outgoing oscillations similar to outgoing sound waves.

A wave, originating from this high-density region at the center of the primordial plasma contains dark matter, baryons and photons (see [1] for a definition of the sub-atomic particles and other items, including the present chapter). This forms a spherical wave of baryons and photons that moves outwards from the central high-density region with a speed about half the speed of light. The dark matter in the wave interacts only by gravitational attraction, and stays at the center of the wave. The photons and baryons move outwards together, and as they progress the photons no longer interact with the baryons and diffuse away. This relieves the pressure on the system, leaving behind a shell of baryonic matter at a fixed radius, usually referred to as the sound horizon. Without the photo-baryon pressure driving the system outwards, only the gravitational force on the baryons remains. Hence the baryons and dark matter (left behind at the center of space-time) form an irregular configuration, which is not the same in all directions (i.e. it is anisotropic) in the densities of matter both at the original site of the anisotropy and in the shell at the sound horizon.

Many such anisotropies were created as density ripples in space and these attract matter. Eventually galaxies form with the same pattern. One would then expect to see a greater number of galaxies that are separated at the sound horizon than at other distances. This particular configuration of matter occurs at each anisotropy in the early universe. Hence the Universe is not composed of one sound ripple, but many overlapping ripples. It is not possible to observe this preferred separation of galaxies on the sound horizon scale by eye, but one can measure it statistically by looking at the separations of a large number of galaxies.

Reference

[1] The Columbia Encyclopedia 2016 6th ed

Sound-Power Flow
A practitioner's handbook for sound intensity

Appendix A

Linear units versus decibels

A.1 Decibels

Decibels have been used for almost a century and have become deeply entrenched in acoustics. However, with the advent of digital signal processing and of routine methods of measuring sound intensity (sound-power flow per unit area) and the use of Pascal units for pressure, their usefulness is now subject to question.

In the United States, common logarithms to the base 10 were used in the early 1900s to measure the relative power loss in electrical transmission lines. The units were called transmission units, TU. In 1923 scientists at Bell Labs renamed the TU the bel. However the bel was too large for use in voice transmission. Hence it was divided by 10 to create the decibel (dB).

In Europe, an equivalent unit was created based on natural logarithms, called the neper (Np), named after John Napier the Scottish mathematician who developed common logarithms at the beginning of the 16th century.

Until the early 1900s, scientific and engineering computations were performed using common logarithms. At that time, Bell Labs had a pre-eminent place in acoustics, having developed vacuum tube amplifiers (1912), condenser microphones (1916), loudspeakers (1918), the first public address system (1921), and the use of modulated light to record sound on moving photographic film (1923). This started the era of mass communication. For the first time, acousticians had the tools to quantify sound. It was felt that decibels should be related to human hearing, based on the wide dynamic range of the human ear. Loudness was considered to be roughly logarithmic based on the Weber–Fechner law, making 0 dB the nominal threshold of hearing.

Since the decibel represents a power ratio, the far-field approximation, valid for plane and spherical progressive waves in a free field, was used to relate sound-power flow (intensity) to sound pressure (the only routinely measurable quantity), thus creating the sound-pressure level, SPL

Table A.1. Comparison between sound-pressure levels in decibels (reference 20 µpa) and the corresponding sound pressure in Pascals.

SPL(dB)	Pascals
0	20 µPa
20	0.2 mPa
40	2 mPa
60	0.02 Pa
80	0.2 Pa
100	2 Pa
120	20 Pa
140	200 Pa
160	2 kPa
180	200 kPa

$$SPL = 20\log 10 \cdot p/p_0 \text{ dB} \qquad (A.1)$$

where p is sound pressure. The reference pressure p_0 is the nominal threshold of hearing, which in air is 20 mPa, and in water is 1 mPa. The pascal (Pa) is the SI unit of pressure in kg ms^{-2}. Relative to atmospheric pressure, it is small (1 Pa = 0.01 millibar), but it is ideally suited to measuring sound pressure, as can be seen from table A.1, which ranges from the nominal threshold of hearing to pressures that can rupture the eardrum [1].

From this table it can be seen that:

1. The overall range in Pascals covers a range equivalent to that of the meter, a quantity in everyday use, which does not use a logarithmic scale.
2. Noise control is concerned roughly with sound in the range from 40 to 90 dB or about 2 mPa to 1 Pa, a range that certainly does not require logarithms.

Further drawbacks of decibel usage

1. Elementary operations of addition and subtraction are needlessly complicated.
2. Inaccuracy is obscured. It is usually assumed by acousticians that an error within 1 dB is acceptable. However, in linear units the error is about 12% for sound pressure and about 24% for sound power, which generally are not acceptable.
3. The difference in reference pressure between air and water can cause confusion. Several years ago there was a major controversy with the US Navy because environmentalists thought that sea mammals were being exposed to sound pressures that were 26 db higher than they actually were. To avoid such confusion, it was proposed to add the reference pressure after the decibel quantity. But of course this increases the complication

involved in the use of decibels. It would have been much simpler just to use the linear quantity.

4. Use of the far-field approximation to derive the sound-pressure level is deceptive, because it seems to imply that it is the only way to determine intensity (sound-power flow) from sound-pressure.

A.2 Octaves and the partitioning of the frequency scale

To find the frequency content of sound, it is necessary to partition the frequency scale. Familiarity with scales such as the piano keyboard, probably led to partitioning into octaves, which is a 2 to 1 frequency ratio. This created intervals that were too large to provide information about frequency content, which then brought about the use of 1/13-octave and 1/12-octave intervals. However, even these are too large in many applications. The human ear can distinguish much smaller differences in frequency.

The advent of electronic computers and digital signal processing in the 1950s–1960s changed the nature of frequency partitioning. For the first time, frequency intervals could be small and evenly distributed. It then became possible to determine frequency content in much greater detail.

A.3 Concluding statement

Strenuous efforts have been made by acousticians to preserve the system of decibels and octaves. However, at this stage, acoustics would appear to be much better off without them. Like Roman numerals they could still be used to create a special aura or other effect.

A.4 Conversion Formulae

Conversion of the sound pressure p to decibels is accomplished using the formula

$$\text{decibels} = 20 \log(p/p_0) \tag{A.1}$$

where p_0 is 20 µPa in air and 1 µPa in water.

Conversion of sound-power flow to decibels is accomplished using the formula,

$$I = 10 \log(i/I_0), \text{ where } I_0 \text{ is } 10 \text{ W/m}^{-2}.$$

Sound-Power Flow
A practitioner's handbook for sound intensity
Appendix B

Physical properties of some gases, liquids, and solids

Table B.1. Gases at $p = 1.013 \times 10$ Pa and $T = 273.15$ K.

Gas	(kg m^{-1})	γ	c (m s^{-1})
Air	1.295	1.402	331.6
Oxygen	1.43	1.40	317.2
Hydrogen	1.41	1.41	1269.5

Table B.2. Liquids at $p = 1.013 \times 10$ Pa and $T = 273.15° + 20$ °K.

Liquid	(kg m^{-1})	B (GPa)	γ	c (m s^{-1})
Fresh water	998	2.18	1.004	1481
Mercury	13 600	25.3	1.13	1450
Turpentine	870	1.07	1.27	1250

Table B.3. Solids.

Solid	(kg m^{-1})	σ	λ (GPa)	μ (GPa)	c (m s^{-1})	c (m s^{-1})
Aluminum	2700	0.33	6.1	2.5	6411	3042
Steel	7700	0.28	13.67	8.3	6100	3749
Brass	8500	0.37	8.11	3.8	4700	2655
Copper	8960	0.35	10.26	4.4	5000	2739
Lucite	1200	0.40	0.56	0.14	2650	1185
Nickel	8900	0.31	13.05	8.0	5850	3606

More complete lists of the acoustic properties of liquids and solids are given by: The Onda Corporation, 592 Waddell Drive, Suite 7, Sunnyvale, California 94089, USA.

doi:10.1088/978-1-6817-4453-7ch27